# Hallucinations

# Oliver Sacks

# Hallucinations

ALFRED A. KNOPF    NEW YORK    TORONTO    2012

THIS IS A BORZOI BOOK
PUBLISHED BY ALFRED A. KNOPF AND
ALFRED A. KNOPF CANADA

www.aaknopf.com
www.randomhouse.ca

Knopf, Borzoi Books, and the colophon are registered
trademarks of Random House, Inc.
Knopf Canada and colophon are trademarks.

Owing to limits of space, permissions acknowledgments
may be found following the index.

Library of Congress Cataloging-in-Publication Data
Sacks, Oliver W.
Hallucinations / Oliver Sacks.—1st American ed.
p. ; cm.
Includes bibliographical references and index.
ISBN 978-0-307-95724-5
I. Title.
[DNLM: 1. Hallucinations. 2. Perceptual Disorders. WM 204]
616.89—dc23      2012002877

Library and Archives Canada Cataloguing in Publication
Sacks, Oliver W.
Hallucinations / Oliver Sacks.
Issued also in electronic format.
ISBN 978-0-307-40217-2
1. Hallucinations and illusions. 2. Perceptual disorders. I. Title.
RC553.H3S23 2012      616.89      C2012-902091-5

Manufactured in the United States of America
First American Edition
www.oliversacks.com

*For Kate*

# Contents

# Introduction

When the word "hallucination" first came into use, in the early sixteenth century, it denoted only "a wandering mind." It was not until the 1830s that Jean-Étienne Esquirol, a French psychiatrist, gave the term its present meaning—prior to that, what we now call hallucinations were referred to simply as "apparitions." Precise definitions of the word "hallucination" still vary considerably, chiefly because it is not always easy to discern where the boundary lies between hallucination, misperception, and illusion. But generally, hallucinations are defined as percepts arising in the absence of any external reality—seeing things or hearing things that are not there.[1]

Perceptions are, to some extent, shareable—you and I can agree that there is a tree; but if I say, "I see a tree there," and you see nothing of the sort, you will regard my "tree" as a hallucination, something concocted by my brain or mind, and imperceptible to you or anyone else. To the hallucinator, though, hallucinations seem very real; they can mimic perception in every respect, starting with the way they are projected into the external world.

---

1. My own favorite definition is that given by William James in his 1890 *Principles of Psychology*: "An hallucination is a strictly sensational form of consciousness, as good and true a sensation as if there were a real object there. The object happens to be not there, that is all." Many other researchers have proposed their own definitions, and Jan Dirk Blom, in his encyclopedic *Dictionary of Hallucinations*, includes dozens of these.

Hallucinations tend to be startling. This is sometimes because of their content—a gigantic spider in the middle of the room or tiny people six inches tall—but, more fundamentally, it is because there is no "consensual validation"; no one else sees what you see, and you realize with a shock that the giant spider or the tiny people must be "in your head."

When you conjure up ordinary images—of a rectangle, or a friend's face, or the Eiffel Tower—the images stay in your head. They are not projected into external space like a hallucination, and they lack the detailed quality of a percept or a hallucination. You actively create such voluntary images and can revise them as you please. In contrast, you are passive and helpless in the face of hallucinations: they happen to you, autonomously—they appear and disappear when they please, not when you please.

There is another mode of hallucination, sometimes called pseudo-hallucination, in which hallucinations are not projected into external space but are seen, so to speak, on the inside of one's eyelids—such hallucinations typically occur in near-sleep states, with closed eyes. But these inner hallucinations have all the other hallmarks of hallucinations: they are involuntary, uncontrollable, and may have preternatural color and detail or bizarre forms and transformations, quite unlike normal visual imagery.

Hallucinations may overlap with misperceptions or illusions. If, looking at someone's face, I see only half a face, this is a misperception. The distinction becomes less clear with more complex situations. If I look at someone standing in front of me and see not a single figure but five identical figures in a row, is this "polyopia" a misperception or a hallucination? If I see someone cross the room from left to right, then see them crossing the room in precisely the same way again and again, is this sort of repetition (a "palinopsia") a perceptual aberration, a hallucination, or both? We tend to speak of such things as misperceptions or illusions if there is something there to begin with—a human figure, for example—whereas hallucinations are conjured out of thin air. But many of my patients experience outright hallucinations, illusions, and complex misperceptions, and sometimes the line between these is difficult to draw.

Though the phenomena of hallucination are probably as old as the human brain, our understanding of them has greatly increased over the last few decades.[2] This new knowledge comes especially from our ability to image the brain and to monitor its electrical and metabolic activities while people are hallucinating. Such techniques, coupled with implanted-electrode studies (in patients with intractable epilepsy who need surgery), have allowed us to define which parts of the brain are responsible for different sorts of hallucinations. For instance, an area in the right inferotemporal cortex normally involved in the perception of faces, if abnormally activated, may cause people to hallucinate faces. There is a corresponding area on the other side of the brain normally employed in reading—the visual word form area in the fusiform gyrus; if this is abnormally stimulated, it may give rise to hallucinations of letters or pseudowords.

---

2. We cannot be certain whether other animals have hallucinations, although "hallucinatory behaviors" have been observed in laboratory animals as well as in natural settings, as Ronald K. Siegel and Murray E. Jarvik described in their review of the subject.

Hallucinations are "positive" phenomena, as opposed to the negative symptoms, the deficits or losses caused by accident or disease, which neurology is classically based on. The phenomenology of hallucinations often points to the brain structures and mechanisms involved and can therefore, potentially, provide more direct insight into the workings of the brain.

Hallucinations have always had an important place in our mental lives and in our culture. Indeed, one must wonder to what extent hallucinatory experiences have given rise to our art, folklore, and even religion. Do the geometric patterns seen in migraine and other conditions prefigure the motifs of Aboriginal art? Did Lilliputian hallucinations (which are not uncommon) give rise to the elves, imps, leprechauns, and fairies in our folklore? Do the terrifying hallucinations of the night-mare, being ridden and suffocated by a malign presence, play a part in generating our concepts of demons and witches or malignant aliens? Do "ecstatic" seizures, such as Dostoevsky had, play a part in generating our sense of the divine? Do out-of-body experiences allow the feeling that one *can* be disembodied? Does the substancelessness of hallucinations encourage a belief in ghosts and spirits? Why has every culture known to us sought and found hallucinogenic drugs and used them, first and foremost, for sacramental purposes?

This is not a new thought—in 1845, Alexandre Brierre de Boismont, in the first systematic medical book on the subject, explored such ideas in a chapter titled "Hallucinations in Relation to Psychology, History, Morality, and Religion." Anthropologists including Weston La Barre and Richard Evans Schultes, among others, have documented the role of hallucinations in societies around the globe.[3] Time has only broadened and deepened our appreciation of the great cultural importance of what might at first seem to be little more than a neurological quirk.

I will say very little in this book about the vast and fascinating realm of dreams (which, one can argue, are hallucinations of a sort), other than to touch on the dreamlike quality of some hallucinations and on the "dreamy states" which occur in some seizures. Some have proposed a continuum of dream states and hallucinations (and this may be especially so with hypnagogic and hypnopompic hallucinations), but, in general, hallucinations are quite unlike dreams.

Hallucinations often seem to have the creativity of imagination, dreams, or fantasy—or the vivid detail and externality of perception. But hallucination is none of these, though it may share some neurophysiological mechanisms with each. Hallucination is a unique and special category of consciousness and mental life.

The hallucinations often experienced by people with schizophrenia also demand a separate consideration, a book of their own, for they cannot be divorced from the often profoundly altered inner life and life circumstances of those with schizophrenia. So I will refer relatively little to schizophrenic hallucinations here, focusing instead on the hallucinations that can occur in "organic" psychoses—the transient psychoses sometimes associated with delirium, epilepsy, drug use, and certain medical conditions.

---

3. La Barre provided an extended review of anthropological perspectives on hallucination in a chapter published in 1975.

M any cultures regard hallucination, like dreams, as a special, privileged state of consciousness—one that is actively sought through spiritual practices, meditation, drugs, or solitude. But in modern Western culture, hallucinations are more often considered to portend madness or something dire happening to the brain—even though the vast majority of hallucinations have no such dark implications. There is great stigma here, and patients are often reluctant to admit to hallucinating, afraid that their friends and even their doctors will think they are losing their minds. I have been very fortunate that, in my own practice and in correspondence with readers (which I think of, in some ways, as an extension of my practice), I have encountered so many people willing to share their experiences. Many of them have expressed the hope that telling their stories will help defuse the often cruel misunderstandings which surround the whole subject.

I think of this book, then, as a sort of natural history or anthology of hallucinations, describing the experiences and impact of hallucinations on those who have them, for the power of hallucinations is only to be understood from first-person accounts.

Some of the chapters that follow are organized by medical categories (blindness, sensory deprivation, narcolepsy, etc.), and others are organized by sensory modality (hearing things, smelling things, etc.). But there is a great deal of overlap and interconnection between these categories, and similar hallucinations may occur in a wide variety of conditions. Here, then, is a sampling which I hope will give a sense of the great range, the varieties, of hallucinatory experience, an essential part of the human condition.

# Hallucinations

# Silent Multitudes:
# Charles Bonnet Syndrome

One day late in November 2006, I got an emergency phone call from a nursing home where I work. One of the residents, Rosalie, a lady in her nineties, had suddenly started seeing things, having odd hallucinations which seemed overwhelmingly real. The nurses had called the psychiatrist in to see her, but they also wondered whether the problem might be something neurological—Alzheimer's, perhaps, or a stroke.

When I arrived and greeted her, I was surprised to realize that Rosalie was completely blind—the nurses had said nothing about this. Though she had not seen anything at all for several years, she was now "seeing" things, right in front of her.

"What sort of things?" I asked.

"People in Eastern dress!" she exclaimed. "In drapes, walking up and down stairs . . . a man who turns towards me and smiles, but he has huge teeth on one side of his mouth. Animals, too. I see this scene with a white building, and it is snowing—a soft snow, it is swirling. I see this horse (not a pretty horse, a drudgery horse) with a harness, dragging snow away . . . but it keeps switching. . . . I see a lot of children; they're walking up and down stairs. They wear bright colors—rose, blue—like Eastern dress." She had been seeing such scenes for several days.

I observed with Rosalie (as with many other patients) that while she was hallucinating, her eyes were open, and even though she could see nothing, her eyes moved here and there, as if looking at an actual scene. It was that which had first caught the nurses' attention. Such looking or scanning does not occur with imagined scenes; most people, when visualizing or concentrating on their internal imagery, tend to close their eyes or else to have an abstracted gaze, looking at nothing in particular. As Colin McGinn brings out in his book *Mindsight*, one does not hope to discover anything surprising or novel in one's own imagery, whereas hallucinations may be full of surprises. They are often much more detailed than imagery, and ask to be inspected and studied.

Her hallucinations, Rosalie said, were more "like a movie" than a dream; and like a movie, they sometimes fascinated her, sometimes bored her ("all that walking up and down, all that Eastern dress"). They came and went, and seemed to have nothing to do with her. The images were silent, and the people she saw seemed to take no notice of her. Apart from their uncanny silence, these figures seemed quite solid and real, though sometimes two-dimensional. But she had never before experienced anything like this, so she could not help wondering: was she losing her mind?

I questioned Rosalie carefully but found nothing suggestive of confusion or delusion. Looking into her eyes with an ophthalmoscope, I could see the devastation of her retinas but nothing else amiss. Neurologically, she was completely normal—a strong-minded old lady, very vigorous for her years. I reassured her about her brain and mind; she seemed, indeed, to be quite sane. I explained to her that hallucinations, strangely, are not uncommon in those with blindness or impaired sight, and that these visions are not "psychiatric" but a reaction of the brain to the loss of eyesight. She had a condition called Charles Bonnet syndrome.

Rosalie digested this and said she was puzzled as to why she had started having hallucinations now, after being blind for several years. But she was very pleased and reassured to be told that her hallucinations represented a recognized condition, one that even had a name. She drew herself up and said, "Tell the nurses—*I* have Charles Bonnet syndrome." Then she asked, "Who was this Charles Bonnet?"

Charles Bonnet was an eighteenth-century Swiss naturalist whose investigations ranged broadly, from entomology to reproduction and regeneration in polyps and other animalcules. When an eye disease made his beloved microscopy impossible, he turned to botany—he did pioneer experiments on photosynthesis—then to psychology, and finally to philosophy. When he heard that his grandfather Charles Lullin had started to have "visions" as his eyesight failed, Bonnet asked him to dictate a full account.

John Locke, in his 1690 *Essay Concerning Human Understanding*, put forward the notion that the mind is a tabula rasa until it receives information from the senses. This "sensationalism," as it was called, was very popular with the philosophes and rationalists of the eighteenth century, including Bonnet. Bonnet also conceived of the brain as "an organ of intricate composition, or rather an assemblage of different organs." These different "organs" all had their own dedicated functions. (Such a modular view of the brain was radical at the time, for the brain was still widely regarded as undifferentiated, uniform in structure and function.) Thus Bonnet attributed his grandfather's hallucinations to continuing activity in what he postulated were visual parts of the brain—an activity drawing on memory now that it could no longer draw on sensation.

Bonnet—who would later experience similar hallucinations when his own eyesight declined—published a brief account of Lullin's experiences in his 1760 *Essai analytique sur les facultés de l'âme*, a book devoted to considering the physiological basis of various senses and mental states, but Lullin's original account, which filled eighteen pages of a notebook, was subsequently lost for nearly 150 years, coming to light only at the beginning of the twentieth century. Douwe Draaisma has recently translated Lullin's account, including it in a detailed history of Charles Bonnet syndrome in his book *Disturbances of the Mind*.[1]

Unlike Rosalie, Lullin still had some eyesight left, and his hallucinations were superimposed on what he saw in the real world. Draaisma summarized Lullin's account:

---

1. Draaisma's book provides not only a vivid account of Bonnet's life and work, but fascinating reconstructions of the lives of a dozen other major figures in neurology whose names are now remembered mostly for the syndromes named after them: Georges Gilles de la Tourette, James Parkinson, Alois Alzheimer, Joseph Capgras, and others.

In February 1758, strange objects had begun to float into his field of vision. It started with something that resembled a blue handkerchief, with a small yellow circle in each corner. . . . The handkerchief followed the movement of his eyes: whether he was looking at a wall, his bed, or a tapestry, the handkerchief blocked out all the ordinary objects in his room. Lullin was perfectly lucid and at no time did he believe that there really was a blue handkerchief floating around. . . .

One day in August two granddaughters came to see him. Lullin was sitting in his armchair opposite the mantelpiece, and his visitors were to his right. From the left, two young men appeared. They were wearing magnificent cloaks, red and grey, and their hats were trimmed with silver. "What handsome gentlemen you've brought with you! Why didn't you tell me they were coming?" But the young ladies swore that they saw no one. Like the handkerchief, the images of the two men dissolved within a few moments. They were followed by many more imaginary visitors in the next few weeks, all of them women; they were beautifully coifed and several of them had a small box on their head. . . .

Somewhat later Lullin was standing at the window when he saw a carriage approaching. It came to a halt at his neighbour's house and, as he watched in amazement, the carriage grew bigger and bigger until it was level with the eaves of the house some thirty feet from the ground, with everything perfectly in proportion. . . . Lullin was amazed by the variety of images he saw: one time it was a swarm of specks that suddenly turned into a flight of pigeons, another time a group of dancing butterflies. Once he saw a rotating wheel floating in the air, the kind you saw in dockside cranes. On a stroll through the

town he stopped to admire an enormous scaffolding, and when
he arrived home he saw the same scaffolding standing in the
living room, but then in miniature, less than a foot high.

As Lullin found, the hallucinations of CBS would come and
go; his lasted for some months and then disappeared for good.

In Rosalie's case, the hallucinations subsided within a few
days, as mysteriously as they had appeared. Almost a year
later, though, I got another phone call from the nurses, telling me that she was "in a terrible state." Rosalie's first words
when she saw me were "All of a sudden, out of a clear blue
sky, the Charles Bonnet has come back with a vengeance."
She described how a few days before, "figures started to walk
around; the room seemed to crowd up. The walls turned into
large gates; hundreds of people started to pour in. The women
were dolled up, had beautiful green hats, gold-trimmed furs,
but the men were terrifying—big, menacing, disreputable,
disheveled, their lips moving as if they were talking."

In that moment, the visions seemed absolutely real to Rosalie. She had all but forgotten that she had Charles Bonnet syndrome. She told me, "I was so frightened that I screamed and
screamed, 'Get them out of my room, open those gates! Get
them out! Then shut the gates!' " She heard a nurse say of her,
"She is not in her right mind."

Now, three days later, Rosalie said to me, "I think I know
what triggered it again." She went on to say that she had had a
highly stressful, exhausting time earlier in the week—a long,
hot journey to see a gastrointestinal specialist on Long Island
and a nasty fall backwards on the way. She arrived back many

hours later, shocked, dehydrated, in a state of near collapse. She was put to bed and fell into a deep sleep. She awoke the next morning to the terrifying visions of people bursting through the walls of her room, which lasted for thirty-six hours. Then she started to feel somewhat better and recovered her insight into what was happening. At that point, she instructed a young volunteer to track down an account of Charles Bonnet syndrome on the internet and to give copies of this to the nursing home staff, so that they would know what had been going on.

Over the next few days, her visions grew much fainter and ceased altogether when she was talking with others or listening to music. Her hallucinations had become "shyer," she said, and now occurred only in the evening, if she sat quietly. I thought of the passage in *Remembrance of Things Past* where Proust speaks of the church bells of Combray, how their sound seemed muted in the daytime, only to be heard when the hubbub and blare of the day had died down.

Charles Bonnet syndrome was considered rare before 1990—there were only a handful of case histories in the medical literature.[2] I thought this strange, for working in

---

2. Or so it would seem. Recently I came across a marvelous 1845 report by Truman Abell, a physician who started to lose his sight in his fifty-ninth year and had become totally blind by 1842, four years later. He described this in an article for the *Boston Medical and Surgical Journal.*

"In this situation," he wrote, "I often dreamed of having my sight restored, and of seeing the most beautiful landscapes. At length these landscapes began to appear in miniature *when awake*: small fields, a few feet square, would appear, clothed with green grass, and other vegetables, some in bloom. These would continue two or three minutes, and then disappear." The landscapes were followed by an immense variety of other "illusions"—Abell did not use the word "hallucinations"—provided by "an internal sight."

old-age homes and nursing homes for over thirty years, I had
seen a number of blind or purblind patients with complex visual
hallucinations of the Charles Bonnet type (just as I had seen
a number of deaf or nearly deaf patients with auditory—and
most often musical—hallucinations). I wondered whether CBS
was actually much commoner than the literature seemed to
indicate. Recent studies have confirmed that this is the case,
although CBS is still little recognized, even by doctors, and
there is much to suggest that many or most cases are overlooked
or misdiagnosed. Robert Teunisse and his colleagues, studying
a population of nearly six hundred elderly patients with visual
problems in Holland, found that almost 15 percent of them had
complex hallucinations—of people, animals, or scenes—and
as many as 80 percent had simple hallucinations—shapes and
colors, sometimes patterns, but not formed images or scenes.

Most cases of CBS probably remain at this elementary level

---

Over the course of several months, his visions increased in complexity. His "silent, but impudent visitors" were sometimes intrusive, with
three or four people who would sit on his bed or "come to my bed-side,
stoop down over me, and look directly into my eyes." (Often his hallucinatory people seemed to acknowledge him, although CBS hallucinations typically do not interact with their hallucinators.) One night, he
reported, "I was threatened to be run over about 10 o'clock by a drove of
oxen; but having my presence of mind, I sat quiet, and with much crowding they all passed without touching me."

Sometimes he saw ranks of thousands of people, splendidly dressed,
forming columns that disappeared into the distance. At one point he saw
"a column at least half a mile wide" of "men on horseback riding towards
the west. . . . They continued to pass for several hours."

"What I have here stated," Abell wrote at the end of his detailed
account, "must appear incredible to those unacquainted with the history
of illusive visions. . . . How far my blindness contributed to produce such
a result, I am not able to say. Never before have I been able to realize the
ancient comparison of the human mind to a microcosm, or universe in
miniature . . . [yet] the whole was confined within the organ of mental
vision, and occupied, perhaps, a space of less than the tenth part of an
inch square."

of simple patterns or colors. Patients who have simple (and perhaps transient or occasional) hallucinations of this type may not take much notice or remember to report them when they visit a doctor. But some people's geometrical hallucinations are more persistent. One old lady with macular degeneration, learning of my interest in such matters, described how in the first two years of her visual impairment, she saw

a big blob of light circling around and then vanishing, followed by a colored flag in sharp focus . . . it looked exactly like the British flag. Where it came from, I do not know. . . . For the last few months I have been seeing hexagons, often hexagons in pink. At first there were also tangled lines inside the hexagons, and other little balls of color, yellow, pink, lavender, and blue. Now there are only black hexagons looking for all the world like bathroom tiles.[3]

While most people with CBS are aware that they are hallucinating (often by the very incongruity of their hallucinations), some hallucinations may be plausible and in context, as with the "handsome gentlemen" accompanying Lullin's granddaughters, and these may, at least initially, be taken as real.[4]

---

3. A particularly good description of hallucinations in CBS ("I See Purple Flowers Everywhere") is provided by Lylas and Marja Mogk in their excellent book *Macular Degeneration*, written for patients with this condition.

4. The reverse may also occur. Robert Teunisse told me how one of his patients, seeing a man hovering outside his nineteenth-floor apartment, assumed this was another one of his hallucinations. When the man waved at him, he did not wave back. The "hallucination" turned out to be his window washer, considerably miffed at not having his friendly wave returned.

With more complex hallucinations, it is typical to see faces, though they are almost never familiar. David Stewart, in an unpublished memoir, described this:

> I had another hallucination. . . . This time it was faces, the most prominent of which was one of a man who might have been a burly ship captain. It wasn't Popeye, but along those lines. The cap he was wearing was blue with a shiny black visor. His face was grey, the cheeks rather chubby, bright eyes and a decidedly bulbous nose. He was no one I had ever seen before. This was not a caricature, and he seemed very much alive, someone I felt I might like to know. He gazed at me with a benign, unblinking, and altogether incurious expression.

The burly ship captain, Stewart noted, appeared as he was listening to an audiobook biography of George Washington, which included a reference to some sailors. He mentioned, too, that he had one hallucination "which nearly replicated a Brueghel painting I once—and only once—observed in Brussels," and another of a coach he thought might have belonged to Samuel Pepys shortly after he read a biography of Pepys.

While some hallucinatory faces, like Stewart's ship captain, seem coherent and plausible, others may be grossly distorted or composed, sometimes, of fragments—a nose, part of a mouth, an eye, a huge head of hair, all juxtaposed in a seemingly haphazard way.

Sometimes people with CBS may hallucinate letters, lines of print, musical notes, numerals, mathematical symbols, or other types of notation. The overall term "text hallucinations" is used for such visions, although for the most part what is seen cannot be read or played and may indeed be nonsensi-

cal. My correspondent Dorothy S. mentioned this as one of her
many CBS hallucinations:

> Then there are the words. They are from no known language,
> some have no vowels, some have too many: "skeeeekkseegsky."
> It is hard for me to capture them as they move swiftly from
> side to side and also advance and retreat. . . . Sometimes I catch
> a glimpse of part of my name, or a version of it: "Doro" or
> "Dorthoy."

Sometimes the hallucinated text has an obvious association with experience, as with one man who wrote to me that
he would see Hebrew letters all over the walls for about six
weeks following Yom Kippur each year. Another man, who
was nearly blind from glaucoma, reported that often he saw
lines of print in balloons, "like the balloons in comic strips,"
though he could not decipher the words. Text hallucinations
are not uncommon; Dominic ffytche, who has seen hundreds
of people with CBS, estimates that about a quarter of them
have text hallucinations of one sort or another.

Marjorie J. wrote to me in 1995 about what she called her
"musical eyes":

> I am a 77-year-old woman with glaucoma damage to mostly
> the lower half of my vision. About two months ago, I started
> to see music, lines, spaces, notes, clefs—in fact written music
> on everything I looked at, but only where the blindness exists.
> I ignored it for a while, but when I was visiting the Seattle Art
> Museum one day and I saw the lines of the explanatory notes as
> music, I knew I was really having some kind of hallucination.
>
> . . . I had been playing the piano and really concentrating

on music prior to the musical hallucinations . . . it was right before my cataract was removed, and I had to concentrate hard to see the notes. Occasionally I'll see crossword puzzle squares . . . but the music does not go away. I've been told the brain refuses to accept the fact that there is visual loss and fills in—with music in my case.

Arthur S., a surgeon who is also a fine amateur pianist, is losing his vision from macular degeneration. In 2007, he started "seeing" musical notation for the first time. Its appearance was extremely realistic, the staves and clefs boldly printed on a white background, "just like a sheet of real music"—and Arthur wondered for a moment whether some part of his brain was now generating his own original music. But when he looked more closely, he realized that the score was unreadable and unplayable. It was inordinately complicated, with four or six staves, impossibly complex chords with six or more notes on a single stem, and horizontal rows of multiple flats and sharps. It was, he said, "a potpourri of musical notation without any meaning." He would see a page of this pseudo-music for a few seconds, and then it would disappear suddenly, replaced by another, equally nonsensical page. These hallucinations were sometimes intrusive, and might cover a page he was trying to read or a letter he was trying to write.

Though Arthur has been unable to read real musical scores for some years, he wonders, as Marjorie did, whether his life-long immersion in music and musical scores might have determined the form of his hallucinations.[5]

5. I have heard from at least a dozen people who, like Arthur and Marjorie, hallucinate musical notation; some of them have eye problems, some parkinsonism, some see music when they have a fever or delirium,

He wonders, too, whether his hallucinations might progress. For about a year before he began to see musical notation, Arthur saw something much simpler: a checkerboard pattern. Will his musical notation be followed by even more complex hallucinations, such as people, faces, or landscapes, as his eyesight declines?

There is clearly a wide array, a whole spectrum, of visual disturbances which can occur when vision is lost or compromised, and originally the term "Charles Bonnet syndrome" was reserved for those whose hallucinations were related to eye disease or other ocular problems. But an essentially similar array of disturbances can also occur when the damage lies not in the eye itself but higher up in the visual system, especially the cortical areas involved in visual perception—the occipital lobes and their projections into the temporal and parietal lobes of the brain—as seems to be the case with Zelda.

---

some see it hypnopompically when they awaken. All but one are amateur musicians who often spend many hours a day studying scores. This very specialized and repetitive sort of visual study is peculiar to musicians. One may read books for hours a day, but one does not usually study print itself in such an intensive way (unless one is a type designer or proofreader, perhaps).

A page of music is far more complex visually than a page of print. With musical notation, one has not just the notes themselves but a very dense set of information contained in symbols for key signatures, clefs, turns, mordents, accents, rests, holds, trills, etc. It seems likely that intensive study and practice of this complex code somehow imprints it in the brain, and should any tendency to hallucination later develop, these "neural imprints" may predispose to hallucinations of musical notation.

And yet people with no particular training or even interest in music may also have hallucinations of musical notation, as Dominic ffytche has pointed out. In a letter to me, he wrote, "although prolonged exposure to music increases the likelihood of musical eyes, it is not a prerequisite."

Zelda was a historian who came to see me in 2008. She told me how her world of strange visual phenomena had started at a theater six years earlier, when the beige curtain in front of the stage suddenly seemed to be covered in red roses—the roses were three-dimensional, thrusting out of the curtain. When she closed her eyes, she still saw the roses. This hallucination lasted for a few minutes and then vanished. She was perplexed and frightened by this, and she went to consult her ophthalmologist, but he found no impairment of vision and no pathological changes in either eye. She saw her internist and cardiologist, but they could not provide any plausible explanation for this episode—or the countless episodes that followed. Finally she had a PET scan of her brain, which showed reduced blood flow in her occipital and parietal lobes, presumably the cause, or at least a possible cause, of her hallucinations.

Zelda has both simple and complex visual hallucinations. The simple ones may appear when she is reading or writing or watching television. One of her physicians asked her to keep a journal of her visions over a three-week period, and in it, she recorded, "As I write this page, it is becoming more and more covered by a pale green and pink lattice. . . . The garage walls, covered in white cinderblock, continually mutate . . . coming to resemble bricks, or clapboard, or being covered with damask, or flowers of different colors. . . . On the upper part of the walls in the hallway, shapes of animals. They were formed by blue dots."

More complex hallucinations—battlements, bridges, viaducts, apartment houses—are especially common when she is being driven in a car (she gave up driving herself after her initial attack, six years ago). Once when she and her husband were driving along a snowy road, she was startled to see brilliant green bushes, their leaves glittering with icicles, to either side of the road. Another day, she saw a rather shocking sight:

> As we drove away from the beauty parlor, I saw what looked like a teenage boy on the front hood of our car, leaning on his arms with his feet up in the air. He stayed there for about five minutes. Even when we turned he stayed on the hood of the car. As we pulled into the restaurant parking lot, he ascended into the air, up against the building, and stayed there until I got out of the car.

At another point, she "saw" one of her great-granddaughters, who rose up, moved to the ceiling, and disappeared. She saw three "witchlike" figures, motionless and hideous, with huge hooked noses, protruding chins, and glaring eyes—these also vanished after a few seconds. Zelda said she had no idea that she had so many hallucinations until she starting keeping a journal; many of them, she thought, would otherwise have been forgotten.

She also spoke of many strange visual experiences which were not quite hallucinations in the sense of being totally invented or generated but seemed to be persistences, repetitions, distortions, or elaborations of visual perceptions. (Charles Lullin had a number of such perceptual disorders, and they are not uncommon in people with CBS.) Some of these were relatively simple; thus, when she looked at me on one occasion, my beard seemed to spread until it covered my entire face and head, and then resumed its proper appearance. Occasionally, looking in a mirror, she might see her own hair rising a foot above her head and have to check with her hand to make sure it was in its usual place.

Sometimes her perceptual changes were more disturbing, as when she encountered her mail carrier in the lobby of her apartment building: "As I looked at her, her nose grew until it was a grotesque figure on her face. After a few minutes, as we stood talking, her face came back to normal."

Zelda would often see objects duplicated or multiplied, and this might create odd difficulties. "Making dinner and eating was quite difficult," she said. "I kept seeing several of each piece of food when they didn't exist. This lasted most of dinner."[6] Visual multiplication like this—polyopia—can take even more dramatic form. Once, in a restaurant, Zelda observed a man in a striped shirt paying at the cash register. As she watched, he split into six or seven identical copies of himself, all wearing striped shirts, all making the same gestures—then concertinaed back into a single person. At other times, her polyopia can be quite frightening or dangerous, as when, sitting in the passenger seat of her car, she saw the road ahead of her split into four identical roads. The car seemed to her to proceed up all four roads simultaneously.[7]

---

6. I was reminded, when she said this, of a case I had heard of in which as a patient ate cherries from a bowl, they were replaced by hallucinatory cherries, a seemingly endless cornucopia of cherries, until, suddenly, the bowl was totally empty. And of another case, of a man with CBS who was picking blackberries. He picked every one he could see; then, to his delight, he saw four more he had missed—but these turned out to be hallucinations.

7. Something about visual movement or "optic flow" seems to be especially provocative of visual hallucinations in people with CBS or other disorders. I met one elderly psychiatrist with macular degeneration who had experienced only a single episode of CBS hallucinations: he was being driven in a car and began to see, on the edges of the parkway, elaborate eighteenth-century gardens which reminded him of Versailles. He enjoyed the experience and found it much more interesting than the ordinary view of the roadside.

Ivy L., who also has macular degeneration, wrote:

Seeing moving pictures even on television may lead to hallucinatory perseverations. Once, watching a television program that showed people descending from a plane, Zelda began to hallucinate minute replicas of the figures, which continued their descent off the screen and down the wooden cabinet of the television console.

Zelda has dozens of these hallucinations or misperceptions every day, and has had them, almost nonstop, for the past six years. And yet she has managed to maintain a very full life, both domestically and professionally—keeping house, entertaining friends, going out with her husband, and completing a new book.

In 2009, one of Zelda's doctors suggested that she take a medication called quetiapine, which can sometimes diminish the severity of hallucinations. To our amazement and especially hers, she became entirely free from hallucinations for more than two years.

In 2011, however, she had heart surgery, and then, on top of this, she broke a kneecap in a fall. Whether it was due to the anxiety and stress of these medical problems, the unpredictable nature of CBS, or the development of tolerance to her medication, she started to have some hallucinations again.

---

As a passenger in cars, I began riding with my eyes closed. Now I often "see" a small, moving travel scene in front of me when my eyes are shut. I "see" open roads and sky, houses, and gardens. I do not "see" any people or vehicles. The scene constantly changes, showing unidentifiable houses in great detail sliding by when the car is in motion. These hallucinations never appear except when I am in a moving car.

(Mrs. L. also reported text hallucinations as part of her CBS, "brief periods when I would 'see' handwriting in huge letters across a large white wall, or the income tax figures imprinted on the drapes. These lasted several years, at intervals.")

Her hallucinations, though, have taken a somewhat more tolerable form. When she is in the car, she said, "I see things, but not people. I see planted fields, flowering, and many forms of medieval buildings. Frequently I see modern buildings change into more historic looking ones. Every experience brings something different."

One of her new hallucinations, she said, "is very difficult to describe. It's a performance! The curtain goes up and 'performers' dance out on the stage—but no people. I see black Hebrew letters dressed in ballet dresses of white. They dance to beautiful music, but I don't know where it comes from. They move the upper parts of the letters like arms and dance on the lower parts so gracefully. They come onstage from right to left."

While the hallucinations of CBS are usually described as pleasant, friendly, diverting, even inspiring, they may occasionally take on a very different character. This happened to Rosalie when a neighbor of hers in the nursing home, Spike, died. Spike was a whimsical, laughter-loving Irishman, and he and Rosalie, both in their nineties, had been close friends for years. "He knew all the old songs," Rosalie remarked; they would sing these together and joke and chatter by the hour. When he died suddenly, Rosalie was devastated. She lost her appetite, withdrew from social activities, and spent more time alone in her room. Her hallucinations returned, but instead of the gaily dressed figures she had seen before, she saw five or six tall men standing around her bed, silent and motionless. They were always dressed in dark brown suits and wore dark hats that shadowed their faces. She could not "see" their eyes, but she felt that they were gazing at her—enigmatically, solemnly. She felt that her bed had become a deathbed and that these ominous figures were harbingers of her own death. They seemed overwhelmingly real to her, and although she knew that if she stretched out a hand it would pass right through them, she could not bring herself to do this.

Rosalie continued to have these visions for three weeks, and then she started to emerge from her melancholy. The somber, silent men in brown disappeared, and her hallucinations started taking place chiefly in the dayroom, a place full of music and talk. They would start with a vision of patterns—quadrangles of pink and blue that seemed to cover the floor and then extend up the walls, finally spreading across the ceiling. The colors of these "tiles," she said, put her in mind of a nursery. And, in accordance with this, she now saw little people a few inches high, like elves or fairies, with little green caps, climbing up the sides of her wheelchair. There were children, too, "picking up pieces of paper from the floor" or climbing hallucinatory stairs in one corner of the room. Rosalie found the children "adorable," although their activities seemed pointless and, as she put it, "silly."

The children and the little people lasted for a couple of weeks, and then they, too, vanished, in the mysterious way that such hallucinations tend to. Though Rosalie misses Spike, she has found other friends in the nursing home, and she is back to her usual routines of chatting and listening to audiobooks and Italian operas. She is rarely alone now, and—coincidentally or not—her hallucinations have, for the time being, disappeared.

If some or all sight is preserved, as with Charles Lullin and Zelda, there may be not only visual hallucinations but various disorders of visual perception: people or objects may appear too large or too small, too near or too far; there may be too little or too much color or depth; misalignment, distortion, or inversion of the image; or problems with motion perception.

If, of course, the person is completely blind, as Rosalie is, then there can only be hallucinations, but these may also show anomalies of color, depth, transparency, motion, scale, and detail. CBS hallucinations are often described as having dazzling, intense color or a fineness and richness of detail far beyond anything one sees with the eyes. There are strong tendencies to repetition and multiplication, so that one may see rows or phalanxes of people, all dressed similarly and making similar motions (some early observers referred to this as "numerosity"). And there is a strong tendency to elaboration: hallucinatory figures often seem to be wearing "exotic dress," rich robes, and strange headgear. Bizarre incongruities often appear, so that a flower may protrude not from someone's hat but from the middle of their face. Hallucinatory figures may be cartoonlike. Faces, in particular, may show grotesque distortions of the teeth or eyes. Some people hallucinate text or music. But by far the commonest hallucinations are the geometrical ones: squares, checkerboards, rhomboids, quadrangles, hexagons, bricks, walls, tiles, tessellations, honeycombs, mosaics. Simplest of all, and perhaps most common, are phosphenes, blobs or clouds of brightness or color, which may or may not differentiate into anything more complex. No single individual has all of these perceptual and hallucinatory phenomena, though some people may have a great range, like Zelda, while others tend to stick to a particular form of hallucination, like Marjorie, with her "musical eyes."

In the last decade or two, Dominic ffytche and his colleagues in London have done pioneering research on the neural basis of visual hallucinations. Based on the detailed reports of dozens of subjects, they developed a taxonomy of hallucinations, including categories like figures with hats, children or small people, landscapes, vehicles, grotesque faces, text, and cartoonlike faces. (This taxonomy is described in a 2000 paper by Santhouse et al.)

With this classification in hand, ffytche went on to do detailed brain-imaging studies in which selected patients with different categories of visual hallucinations were asked to signal the beginning and end of their hallucinations while being scanned.

There was, as ffytche et al. wrote in a 1998 paper, "a striking correspondence" between the particular hallucinatory experiences of each patient and the particular portions of the ventral visual pathway in the visual cortex which were activated. Hallucinations of faces, of color, of textures, and of objects, for example, each activated particular areas known to be involved in specific visual functions. When there were colored hallucinations, there was activation of areas in the visual cortex associated with color construction; when there were facial hallucinations of a sketchlike or cartoonlike character, there was activation in the fusiform gyrus. Visions of deformed or dismembered faces or grotesque faces with exaggerated eyes or teeth were associated with heightened activity in the superior temporal sulcus, an area specialized for the representation of eyes, teeth, and other parts of the face. Text hallucinations are associated with abnormal activation in the visual word form area, a highly specialized area in the left hemisphere.

Ffytche et al. observed, moreover, a clear distinction between normal visual imagination and actual hallucination—thus, imagining a colored object, for example, did not activate the V4 area, while a colored hallucination did. Such findings confirm that, not only subjectively but physiologically, hallucinations are unlike imagination and much more like perceptions. Writing of hallucinations in 1760, Bonnet said, "The mind would not be able to tell apart vision from reality." The work of ffytche and his colleagues shows that the brain does not distinguish them, either.

There had never before been direct evidence of such a correlation between the contents of a hallucination and the particular areas of cortex activated. We have long known, from observation of people with specific injuries or strokes, that different aspects of visual perception (color perception, face recognition, movement perception, etc.) depend on highly specialized areas of the brain. Thus, for example, damage to a tiny area of the visual cortex called V4 may knock out color perception but nothing else. Ffytche's work is the first to confirm that hallucinations make use of the same visual areas and pathways as perception itself. (Ffytche has emphasized more recently, in papers on the "hodology" of hallucinations, that attributing hallucinations, or any cerebral function, to specific brain regions has its limitations, and that one must pay equal attention to the connections between these areas.)[8]

---

8. Such correlations involve sizable regions of the brain; they are at a macro level. Correlations on a micro level, at least for elementary geometric hallucinations, have been proposed by William Burke, a neurophysiologist who has experienced such hallucinations himself, due to macular "holes" in both eyes. He has been able to estimate the visual angles subtended by specific hallucinations and to extrapolate these into cortical distances. He concludes that the separation of his brickwork hallucinations corresponds to the separation of the physiologically active "stripes" in the V2 part of the visual cortex, while the separation of the dots he hallucinates corresponds to that of the "blobs" in the primary

But while there are neurologically determined categories of visual hallucination, there may be personal and cultural determinants, too. No one can have hallucinations of musical notation or numbers or letters, for example, if they have not actually seen these at some point in real life. Thus experience and memory may influence both imagery and hallucination—but with CBS, memories are not hallucinated in full or literal form. When people with CBS hallucinate people or places, they are almost never recognizable people or places, only plausible or invented ones. CBS hallucinations give one the impression that, at some lower level, in the early visual system, there is a categorical dictionary of images or part images—of generic "noses," for example, or "headwear" or "birds," rather than of particular noses or headwear or birds. These are, so to speak, the visual ingredients called upon and used in the recognition and representation of complex scenes—elements or building blocks which are purely visual, without context or correlation with other senses, without emotion or particular associations of place or time. (Some researchers have called them "proto-objects" or "proto-images.") In this way, CBS images seem more raw, more obviously neurological, not personal like those of imagination or recollection.

Hallucinations of text or musical scores are intriguing in this regard, for although they initially look like real music or text, they quickly reveal themselves to be unreadable, in the sense that they have no shape, no tune, no syntax or grammar. Although Arthur S. at first thought he might be able to play

---

visual cortex. Burke hypothesizes that with diminished input from his damaged maculas, there is diminished activity in the macular cortex, releasing spontaneous activity in the cortical stripes and blobs that give rise to hallucinations.

his hallucinatory musical scores, he soon realized that he was seeing "a potpourri of musical notation, without any meaning." Similarly, text hallucinations lack meaning; they may, on closer inspection, not even be actual letters but letter-like runes.

We know (from studies by ffytche and his colleagues) that text hallucinations go with hyperactivity in the visual word form area; there are probably analogous (though more widespread) activations with hallucinations of musical notation, though these have yet to be "caught" on fMRI. In the normal process of reading text or scores, what is initially deciphered in the early visual system goes on to higher levels where it acquires syntactical structure and meaning. But in hallucinations of text or scores, caused by anarchical hyperactivity in the early visual system, letters, proto-letters or musical notation appear without the normal constraints of syntax and meaning—providing a window into both the powers and the limitations of the early visual system.

Arthur S. saw musical notation of fanciful elaboration, far more ornate than any real score. CBS hallucinations are often fanciful or fantastical. Why should Rosalie, a blind old woman in the Bronx, see figures in "Eastern dress"? This strong disposition to the exotic, for reasons we do not yet understand, is characteristic of CBS, and it would be fascinating to see whether this varies in different cultures. These strange, sometimes surreal images, of boxes or birds perched atop people's heads or flowers coming out of their cheeks, make one wonder whether what is occurring is a sort of neurological mistake, a simultaneous activation of different brain areas, producing an involuntary, incongruous collision or conflation.

The images of CBS are more stereotyped than those of

dreams and at the same time less intelligible, less meaning-
ful. When Lullin's notebook, lost for a century and a half,
resurfaced and was published in a psychology journal in 1902
(just two years after Freud's *Interpretation of Dreams*), some
wondered whether the hallucinations of CBS might afford, as
Freud felt dreams did, "a royal road" to the unconscious. But
attempts at "interpreting" CBS hallucinations in this sense
bore no fruit. People with CBS had their own psychodynamics,
of course, like everyone else, but it became apparent that little beyond the obvious was to be gained from analyzing their
hallucinations. A religious person might hallucinate praying
hands, among other things, or a musician might hallucinate
musical notation, but these images scarcely yielded insights
into the unconscious wishes, needs, or conflicts of the person.

Dreams are neurological as well as psychological phenomena, but very unlike CBS hallucinations. Dreamers are wholly
enveloped in their dreams, and usually active participants in
them, whereas people with CBS retain their normal, critical
waking consciousness. CBS hallucinations, even though they
are projected into external space, are marked by a lack of interaction; they are always silent and neutral—they rarely convey
or evoke any emotion. They are confined to the visual, without
sound, smell, or tactile sensation. They are remote, like images
on a cinema screen in a theater one has chanced to walk into.
The theater is in one's own mind, and yet the hallucinations
seem to have little to do with one in any deeply personal sense.

One of the defining characteristics of Charles Bonnet hallucinations is the preservation of insight, the realization
that a hallucination is not real. People with CBS are occasion-

ally deceived by a hallucination, especially if it is plausible or contextually appropriate. But such mistakes are quickly realized to be such, and insight is restored. The hallucinations of CBS almost never lead to persistent false ideas or delusions.

The ability to evaluate one's perceptions or hallucinations, however, may be compromised if there are other underlying problems in the brain, especially those which impair the frontal lobes, since the frontal lobes are the seat of judgment and self-evaluation. This may happen transiently, for example, with a stroke or head injury; fever or delirium; various medications, toxins, or metabolic imbalances; dehydration or lack of sleep. In such cases, insight will return as soon as cerebral function returns to normal. But if there is an ongoing dementia, like Alzheimer's or Lewy body disease, there may be less and less ability to recognize hallucinations as such—which, in turn, may lead to frightening delusions and psychoses.

Marlon S., in his late seventies, has progressive glaucoma and some mild dementia. He has been unable to read for the last twenty years, and for the last five years has been virtually blind. He is a devout Christian and still works as a lay minister in prisons, as he has done for the last thirty years. He lives alone in an apartment, but he leads a very active social life. He goes out each day, either with one of his children or with a home attendant, to family occasions or to the senior center, where there may be games, dancing, going out to restaurants, and other activities.

Although he is blind, Marlon seems to inhabit a world that is very visual and sometimes very strange. He tells me that he often "sees" his surroundings—he has lived most of his life in the Bronx, but what he sees is an ugly, desolate version of the Bronx (he describes it as "shabby, old, much older than me"), and this may give him a feeling of disorientation. He "sees" his apartment, but he can easily get lost or confused. Sometimes, he says, the apartment gets "as big as a Greyhound bus terminal," and at other times it contracts, becoming "as skinny as a railroad apartment." In general, the hallucinated apartment looks dilapidated and chaotic: "My whole house is a wreck, looks like the Third World . . . then it looks regular." (The only time his apartment actually *is* a mess, his daughter told me, is when Marlon, thinking that he is "blockaded" by the furniture, starts rearranging it, pushing things to and fro.)

His hallucinations started about five years ago and were at first benign. "In the beginning," he told me, "I saw a lot of animals." They were followed by hallucinations of children—multitudes of them, just as there would always be multitudes of animals. "All of a sudden," Marlon remembered, "I see all these kids come in, they were walking all around; I thought they were regular kids." The children were silent but "talked with their hands"; they seemed unconscious of him and "did their own thing"—walking around, playing. He was startled when he found that no one else saw them. It was only then that he realized that his "eyes were playing tricks" on him.

Marlon enjoys listening to talk shows, gospel, and jazz on the radio, and when he does so, he may find his sitting room crowded with hallucinatory people who are also listening. Sometimes their mouths move as if they are speaking or singing along with the radio. These visions are not unpleasant, and they seem to provide a sort of hallucinatory comfort. It is a social scene, which he enjoys.[9]

---

9. I have heard similar descriptions from other people who have both CBS and some dementia. Janet B. likes to listen to audiobooks and sometimes finds herself joined by a hallucinatory group of fellow listeners.

In the last two years, Marlon has also started to see a mysterious man who always wears a brown leather coat, green pants, and a Stetson hat. Marlon has no idea who it is but feels that this man has a special message or meaning, though what the message or meaning is eludes him. He sees this figure at a distance, never close up. The man seems to float through the air rather than walk, and his figure can become enormous, "as tall as a house." Marlon has also spotted a small, sinister trio of men, "like FBI, a long way off. . . . They look real, real ugly and bad." Marlon believes in angels and devils, he tells me, and he feels that these men are evil. He has started to suspect that he is under surveillance by them.

Many people with mild cognitive impairment may be organized and oriented during the daytime—this is the case with Marlon, especially when he is at the senior center or at a church social, actively engaged with other people. But as

---

They listen intently but never speak, do not respond to her questions, and seem unaware of her presence. At first, Janet realized that they were hallucinatory, but later, as her dementia advanced, she insisted that they were real. Once when her daughter was visiting and said, "Mom, there's no one here," she got angry and chased her daughter out.

A more complex delusional overlay occurred while she was listening to a favorite show on television. It seemed to Janet that the television crew had decided to use her apartment, and that it was set up with cables and cameras, that the show was actually being filmed at that moment in front of her. Her daughter happened to telephone her during the show, and Janet whispered, "I have to be quiet—they're filming." When her daughter arrived an hour later, Janet insisted that there were still cables all over the floor, adding, "Don't you see that woman?"

Even though Janet was convinced of the reality of these hallucinations, they were entirely visual. People pointed, gestured, mouthed, but made no sound. Nor did she have any sense of personal involvement; she found herself in the midst of strange happenings, yet they seemed to have nothing to do with her. In this way they retained the typical character of CBS hallucinations, even though she insisted that they were real.

evening comes, there may be a "sundowning" syndrome, and fears and confusions start to proliferate.

Generally, in the daytime, Marlon's hallucinatory figures deceive him briefly, for a minute or two, before he realizes they are figments. But late in the day, his insight breaks down, and he feels his threatening visitors as real. At night, when he finds "intruders" in his apartment, he is terrified—even though they seem uninterested in him. Many of them look "like criminals" and wear prison garb; sometimes they are "smoking Pall Malls." One night one of his intruders was carrying a bloodstained knife, and Marlon yelled out, "Get out of here, in the name of the blood of Jesus!" On another occasion one of the apparitions left "under the door," slipping away like a liquid or vapor. Marlon has ascertained that these figures are "like ghosts, not solid," and that his arm will go right through them. Nevertheless, they *seem* quite real. He can laugh about this as we talk, but it is clear that he can be quite terrified and deluded when he is alone with his intruders in the middle of the night.

People with CBS have, at least in part, lost the primary visual world, the world of perception. But they have gained, if only in an inchoate and fitful way, a world of hallucinations, a secondary visual world. The role CBS may play in an individual's life varies enormously, depending on the sort of hallucinations that occur, how often they occur, and whether they are contextually appropriate, or frightening, or comforting, even inspiring. There are, at one extreme, those who may have had only a single hallucinatory experience in their life; others may have had hallucinations, on and off, for years. Sometimes hallucinations can be distracting—seeing patterns or webs over everything, not knowing whether the food on one's plate is real or hallucinatory. Some hallucinations are manifestly unpleasant, especially those that involve deformed or dismembered faces. A few are dangerous: Zelda, for instance, does not dare drive, since she may see the road suddenly bifurcate or people jumping on the hood of her car.

For the most part, however, the hallucinations of CBS are unthreatening and, once accommodated to, mildly diverting. David Stewart speaks of his hallucinations as being "altogether friendly," and he imagines his eyes saying, "Sorry to have let you down. We recognize that blindness is no fun, so we've organized this small syndrome, a sort of coda to your sighted life. It's not much, but it's the best we can manage."

Charles Lullin, too, enjoyed his hallucinations and would sometimes go into a quiet room for a brief hallucinatory break. "His mind makes merry with the images," Bonnet wrote of his grandfather. "His brain is a theatre where the stage machinery puts on performances which are all the more amazing because they are unexpected."

Sometimes the hallucinations of Charles Bonnet syndrome can inspire. Virginia Hamilton Adair wrote poetry as a young woman, publishing in the *Atlantic Monthly* and the *New Republic*. She continued to write poems during her career as a scholar and professor of English in California, but these, for the most part, remained unpublished. It was not until she was eighty-three and completely blind from glaucoma that she published her first book of poetry, the acclaimed *Ants on the Melon*. Two further collections followed, and in these new poems she made frequent reference to the Charles Bonnet hallucinations that now visited her regularly, the visions given to her by "the angel of hallucinations," as she put it.

Adair and, later, her editor sent me extracts from the journal she kept in the last years of her life. They were full of descriptions she dictated of her hallucinations as they occurred, including this:

I am maneuvered into a delightfully soft chair. I sink, submerged as usual in shades of night . . . the sea of clouds at my feet clears, revealing a field of grain, and standing about it a small flock of fowl, not two alike, in somber plumage: a miniature peacock, very slender, with its little crest and unfurled tail feathers, some plumper specimens, and a shore bird on long stems, etc. Now it appears that several are wearing shoes, and among them a bird with four feet. One expects more color among a flock of birds, even in the hallucinations of the blind. . . . The birds have turned into little men and women in medieval attire, all strolling away from me. I see only their backs, short tunics, tights or leggings, shawls or kerchiefs. . . . Opening my eyes on the smoke screen of my room I am treated to stabs of sapphire, bags of rubies scattering across the night, a legless vaquero in a checked shirt stuck on the back of a small steer, bucking, the orange velvet head of a bear decapitated, poor thing, by the guard of the Yellowstone Hotel garbage pit. The familiar milkman invaded the scene in his azure cart with the golden horse; he joined us a few days ago out of some forgotten book of nursery rhymes or the back of a Depression cereal box. . . . But the magic lantern show of colored oddities has faded and I am back in black-wall country without form or substance . . . where I landed as the lights went out.

# The Prisoner's Cinema: Sensory Deprivation

The brain needs not only perceptual input but perceptual *change*, and the absence of change may cause not only lapses of arousal and attention but perceptual aberrations as well. Whether darkness and solitude is sought out by holy men in caves or forced upon prisoners in lightless dungeons, the deprivation of normal visual input can stimulate the inner eye instead, producing dreams, vivid imaginings, or hallucinations. There is even a special term for the trains of brilliantly colored and varied hallucinations which come to console or torment those kept in isolation or darkness: "the prisoner's cinema."

Total visual deprivation is not necessary to produce hallucinations—visual monotony can have much the same effect. Thus sailors have long reported seeing things (and perhaps hearing them, too) when they spent days gazing at a becalmed sea. It is similar for travelers riding across a featureless desert or polar explorers in a vast, unvarying icescape.

Soon after World War II, such visions were recognized as a special hazard for high-altitude pilots flying for hours in an empty sky, and it is a danger for long-distance truckers focused for hours on an endless road. Pilots and truckers, those who monitor radar screens for hours on end—anyone with a visually monotonous task is susceptible to hallucinations. (Similarly, auditory monotony may lead to auditory hallucinations.)

In the early 1950s, researchers in Donald Hebb's laboratory at McGill University designed the first experimental study of prolonged perceptual isolation, as they called it (the term "sensory deprivation" became popular later). William Bexton and his colleagues investigated this with fourteen college students immured in soundproof cubicles for several days (except for brief time out for eating and going to the toilet), wearing gloves and cardboard cuffs to reduce tactile sensation and translucent goggles which allowed only a perception of light and dark.

At first the test subjects tended to fall asleep, but then, on awakening, they became bored and craved stimulation—stimulation not available from the impoverished and monotonous environment they were in. And at this point, self-stimulation of various sorts began: mental games, counting, fantasies, and, sooner or later, visual hallucinations—usually a "march" of hallucinations from simple to complex, as Bexton et al. described:

In the simplest form the visual field, with the eyes closed, changed from dark to light colour; next in complexity were dots of light, lines, or simple geometrical patterns. All 14 subjects reported such imagery, and said it was a new experience to them. Still more complex forms consisted in "wall-paper patterns," reported by 11 subjects, and isolated figures or objects, without background (e.g., a row of little yellow men with black caps on and their mouths open; a German helmet), reported by seven subjects. Finally, there were integrated scenes (e.g., a procession of squirrels with sacks over their shoulders marching "purposefully" across a snow field and out of the field of "vision"; prehistoric animals walking about in the jungle). Three of the 14 subjects reported such scenes, frequently including dreamlike distortions, with the figures often being described as "like cartoons."

While these images first appeared as if projected onto a flat screen, after a time they became "compellingly three-dimensional" for some of the subjects, and parts of a scene might become inverted or pivot from side to side.

After being initially startled, the subjects tended to find their hallucinations amusing, interesting, or sometimes irritating ("their vividness interfered with sleep") but without any "meaning." The hallucinations seemed external, proceeding autonomously, with little relevance or reference to the individual or situation. The hallucinations usually disappeared when the subjects were asked to do complex tasks like multiplying three-figure numbers, but not if they were merely exercising or talking to the researchers. The McGill researchers reported, as many others have, auditory and kinesthetic hallucinations as well as visual ones.

This and subsequent studies aroused enormous interest in the scientific community, and both scientific and popular efforts were made to duplicate the results. In a 1961 paper, John Zubek and his colleagues reported, in addition to hallucination, a change in visual imagery in many of their subjects:

At various intervals . . . the subjects were asked to imagine or visualize certain familiar scenes, for example, lakes, countryside, the inside of their homes, and so forth. The majority of the subjects reported that the images which they conjured up were of unusual vividness, were usually characterized by bright colours, and had considerable detail. All these subjects were unanimous in their opinion that their images were more vivid than anything they had previously experienced. Several subjects who normally had great difficulty visualizing scenes could now visualize them almost instantly with great vividness. . . . One subject . . . could visualize faces of former associates of a few years back with almost picture-like clarity, a thing which he was never able to do previously. This phenomenon usually appeared during the second or third day and, in general, became more pronounced with time.

Such visual heightenings—whether due to disease, deprivation, or drugs—can take the form of enhanced visual imagery or hallucination or both.

In the early 1960s, sensory deprivation tanks were designed to intensify the effect of isolation by floating the body in a darkened tank of warm water, which removed not only any sense of bodily contact with the environment but also the proprioceptive sense of the body's position and even its existence. Such immersion chambers could produce "altered states" much more profound than those described in the original experiments. At the time, such sensory deprivation tanks were sought out as avidly as (and sometimes combined with) "consciousness-expanding" drugs, which were more widely available then.[1]

There was a great deal of research on sensory deprivation in the 1950s and 1960s (a 1969 book edited by Zubek entitled *Sensory Deprivation: Fifteen Years of Research* listed thirteen hundred references)—but then scientific interest, like popular interest, started to peter out, and there was relatively little research until the recent work of Alvaro Pascual-Leone and his colleagues (Merabet et al.), who designed a study to isolate the effects of pure visual deprivation. Their subjects, though blindfolded, were able to move around freely and "watch" TV, listen to music, walk outside, and talk to others. They experienced none of the somnolence, boredom, or restlessness the earlier test subjects had shown. They were alert and active during the daytime, when they carried tape recorders so they could take immediate note of any hallucinations. They enjoyed calm, restful sleep at night, and each morning they dictated what they could remember of their dreams—dreams that did not seem significantly altered by their being blindfolded.

The blindfolds, which allowed the subjects to close or move their eyes, were worn continuously for ninety-six hours. Ten of the thirteen subjects experienced hallucinations, sometimes during the first hours of blindfolding, but always by the second day, whether their eyes were open or not.

Typically the hallucinations would appear suddenly and

---

1. While the romantic use of sensory deprivation, as that of vision-producing drugs, has diminished since the 1960s, its political use is still horrifyingly common in the treatment of prisoners. In a 1984 paper on "hostage hallucinations," Ronald K. Siegel pointed out that such hallucinations can be magnified sometimes to madness, especially when combined with social isolation, sleep deprivation, hunger, thirst, torture, or the threat of death.

spontaneously, then disappear just as suddenly after seconds or minutes—although in one subject, hallucination became almost continuous by the third day. The subjects reported a range from simple hallucinations (flashing lights, phosphenes, geometrical patterns) to complex ones (figures, faces, hands, animals, buildings, and landscapes). In general, the hallucinations appeared full-fledged, without warning—they never seemed to be built up slowly, piecemeal, like voluntary imagery or recall. For the most part, the hallucinations aroused little emotion and were regarded as "amusing." Two subjects had hallucinations which correlated with their own movements and actions: "I have the sensation that I can see my hands and my arms moving when I move them and leaving an illuminated trail," said one subject. "I felt like I was seeing the pitcher while I was pouring the water," said another.

Several subjects spoke of the brilliance and colors of their hallucinations; one described "resplendent peacock feathers and buildings." Another saw sunsets almost too bright to bear and luminous landscapes of extraordinary beauty, "much prettier, I think, than anything I have ever seen. I really wish I could paint."

Several mentioned spontaneous changes in their hallucinations; for one subject, a butterfly became a sunset, which changed to an otter and, finally, a flower. None of the subjects had any voluntary control over their hallucinations, which seemed to have "a mind" or "a will" of their own.

No hallucinations were experienced when subjects were engaged in challenging sensory activity of another mode, such as listening to television or music, talking, or even attempting to learn Braille. (The study was concerned not only with hallucinations but with the power of blindfolding to improve and heighten tactile skills and the ability to conceive of space and the world around one in nonvisual terms.)

Merabet et al. felt that the hallucinations reported by their subjects were entirely comparable with those experienced by patients with Charles Bonnet syndrome, and their results suggested to them that visual deprivation alone could be a sufficient cause for CBS.[2]

But what exactly is going on in the brains of such experimental subjects—or in the brains of pilots who crash in cloudless blue skies, or truckers who see phantoms on an empty road, or prisoners watching their enforced "cinema" in darkness?

With the advent of functional brain imaging in the 1990s it became possible to visualize, at least in gross terms, how the brain might respond to sensory deprivation—and, if one was lucky (hallucinations are notoriously fickle, and the inside of an fMRI machine is not an ideal place for delicate sensory experiences), one might even catch the neural correlates of a fugitive hallucination. One such study, by Babak Boroojerdi and his colleagues, showed an increase in the excitability of the visual cortex when subjects were visually deprived, a change that occurred within minutes. Another group of researchers, in the neuroscience lab led by Wolf Singer, studied a single subject, a visual artist with excellent powers of visual imagery (an article on this by Sireteanu et al. was published in 2008).

2. There may be severe visual impairment or complete blindness without a hint of CBS, and this might seem to imply that visual deprivation alone is not a sufficient cause for it. But we are still ignorant as to why some people with visual problems get CBS and others do not.

The subject was blindfolded for twenty-two days and spent several sessions in an fMRI machine, where she was able to indicate the exact times her hallucinations appeared and disappeared. The fMRI showed activations in her visual system, both in the occipital cortex and in the inferotemporal cortex, in precise coincidence with her hallucinations. (When, by contrast, she was asked to recall or imagine the hallucinations using her powers of visual imagery, there was, additionally, a good deal of activation in the executive areas of the brain, in the prefrontal cortex—areas that had been relatively inactive when she was merely hallucinating.) This made it clear that, at a physiological level, visual imagery differs radically from visual hallucination. Unlike the top-down process of voluntary visual imagery, hallucination is the result of a direct, bottom-up activation of regions in the ventral visual pathway, regions rendered hyperexcitable by a lack of normal sensory input.

The deafferentation tanks used in the 1960s produced not only visual deprivation but every other sort of deprivation: of hearing, touch, proprioception, movement, and vestibular sensation, as well as, to varying degrees, deprivation of sleep and social contact—any of which may in themselves lead to hallucinations.

Hallucinations engendered by immobility, whether from motor system disease or external constraints, were frequently seen when polio was rampant. The worst afflicted, unable even to breathe by themselves, lay motionless in coffinlike "iron lungs" and would often hallucinate, as Herbert Leiderman and his colleagues described in a 1958 article. The immobility produced by other paralyzing diseases—or even splints and casts for broken bones—may likewise provoke hallucinations. Most commonly these are corporeal hallucinations, in which limbs may seem to be absent, distorted, misaligned, or multiplied; but voices, visual hallucinations, and even full-blown psychoses have been reported, too. I saw this especially with my post-encephalitic patients, many of whom were, in effect, enclosed in immoveable parkinsonism and catatonia.

Sleep deprivation beyond a few days leads to hallucination, and so may dream deprivation, even with otherwise normal sleep. When this is combined with exhaustion or extreme physical stress, it can be an even more potent source of hallucinations. Ray P., a triathlete, described one example:

> Once, I was competing in the Ironman Triathlon in Hawaii. I was not having a good race, I was overheated and dehydrated—miserable. Three miles into the marathon portion of the race, I saw my wife and my mom standing on the side of the road. I ran over to them to say I would be late to the finish line, but when I reached them and began telling my tale of woe, two complete strangers who did not even remotely resemble my wife and mother looked back at me.
>
> The Hawaiian Ironman Triathlon, with its extreme temperatures and long hours of monotony under grueling conditions, can provide an athlete with a fertile venue for hallucination, much the same as the vision quest rites of passage of Native Americans. I have seen Madame Pele, the Hawaiian Volcano and Fire Goddess, at least once out there in the lava fields.

Michael Shermer has spent much of his life debunking the paranormal; he is a historian of science and the director of the

Skeptics Society. In his book *The Believing Brain*, he provides other examples of hallucinations in marathon athletes, like those of the mushers competing in the Iditarod dogsled race:

> Mushers go for 9–14 days on minimum sleep, are alone except for their dogs, rarely see other competitors, and hallucinate horses, trains, UFOs, invisible airplanes, orchestras, strange animals, voices without people, and occasionally phantom people on the side of the trail or imaginary friends. . . . A musher named Joe Garnie became convinced that a man was riding in his sled bag, so he politely asked the man to leave, but when he didn't move Garnie tapped him on the shoulder and insisted he depart his sled, and when the stranger refused Garnie swatted him.

Shermer, an endurance athlete himself, had an uncanny experience while competing in a grueling bike marathon, which he later described in his *Scientific American* column:

> In the wee hours of the morning of August 8, 1983, while I was traveling along a lonely rural highway approaching Haigler, Neb., a large craft with bright lights overtook me and forced me to the side of the road. Alien beings exited the craft and abducted me for 90 minutes, after which time I found myself back on the road with no memory of what transpired inside the ship. . . . My abduction experience was triggered by sleep deprivation and physical exhaustion. I had just ridden a bicycle 83 straight hours and 1,259 miles in the opening days of the . . . transcontinental Race Across America. I was sleepily weaving down the road when my support motor home flashed its high beams and pulled alongside, and my crew entreated me to take a sleep break. At that moment a distant memory

of the 1960s television series "The Invaders" was inculcated into my waking dream. In the series, alien beings were taking over the earth by replicating actual people but, inexplicably, retained a stiff little finger. Suddenly the members of my support crew were transmogrified into aliens. I stared intensely at their fingers and grilled them on both technical and personal matters.

After a nap, Shermer recognized this as a hallucination, but at the time it seemed completely real.

3

# A Few Nanograms of Wine: Hallucinatory Smells

The ability to imagine smells, in normal circumstances, is not that common—most people cannot imagine smells with any vividness, even though they may be very good at imagining sights or sounds. It is an uncommon gift, as Gordon C. wrote to me in 2011:

> Smelling objects that are not visible seems to have been a part of my life for as long as I can remember. . . . If, for instance, I think for a few minutes about my long dead grandmother, I can almost immediately recall with near perfect sensory awareness the powder that she always used. If I'm writing to someone about lilacs, or any specific flowering plant, my olfactory senses produce that fragrance. This is not to say that merely writing the word "roses" produces the scent; I have to recall a specific instance connected with a rose, or whatever, in order to produce the effect. I always considered this ability to be quite

natural, and it wasn't until adolescence that I discovered that it was not normal for everyone. Now I consider it a wonderful gift of my specific brain.

Most of us, in contrast, have difficulty summoning smells to mind, even with strong suggestion. And it may be oddly difficult to know whether a smell is real or not. Once I revisited the house where I grew up and where my family lived for sixty years. The house had been sold to the British Association of Psychotherapists in 1990, and what used to be our dining room had been turned into an office. When I entered this room on a visit in 1995, I immediately and strongly smelled the kosher red wine which used to be kept in a wooden sideboard next to the dining table and drunk with Kiddush on the Sabbath. Was I just imagining the smell, assisted by these once intensely familiar, beloved surroundings and nearly sixty years of memory and association? Or could a few nanograms of wine have survived all of the repainting and renovation? Smells can be oddly persistent, and I am not sure whether my experience should be called a heightened perception, a hallucination, a memory, or some combination of all these.

My father had an acute sense of smell as a young man, and like all doctors of his generation, he depended on it when seeing patients. He could detect the smell of diabetic urine or of a putrid lung abscess as soon as he entered a patient's house. A series of sinus infections in middle age blunted his sense of smell, and he could no longer rely on his nose as a diagnostic tool. But he was fortunate that he did not lose his sense of smell entirely, for total loss of the sense of smell—anosmia, which affects perhaps as many as 5 percent of people—causes many problems. People with anosmia cannot smell gas, smoke, or rancid food; they may be beset by social anxiety, not knowing whether they themselves smell of something rank. They cannot enjoy the good smells of the world, either, and they cannot enjoy many of the subtler flavors of food (for most of these depend equally on smell).[1]

I wrote about one anosmic patient in *The Man Who Mistook His Wife for a Hat*. He had suddenly lost all sense of smell, as the result of a head injury. (The long olfactory tracts are easily sheared as they cross the base of the skull, so loss of smell can be caused by a relatively mild head injury.) This man had never given much conscious thought to the sense of smell, but once he lost it, he found his life radically poorer. He missed the smell of people, of books, of the city, of springtime. He hoped against hope that the lost sense would return. And, indeed, it seemed to come back some months later when, to his surprise and delight, he smelled his morning coffee as it was brewing. Tentatively, he tried his pipe, abandoned for many months, and caught a whiff of his favorite aromatic tobacco. He returned, excitedly, to his neurologist, but after careful testing, he was told that there was not a trace of recovery. Clearly, though, he was having an olfactory experience of some sort, and I could only think that his power to imagine smells, at least in situations charged with memories and associations, had been enhanced by his anosmia, perhaps as the power to visualize may be enhanced in some who have lost their sight.

---

1. Molly Birnbaum, an aspiring chef who became anosmic after being struck by a car, has described the anosmic's predicament eloquently in her memoir *Season to Taste*.

The heightened sensitivity of sensory systems when they
have lost their normal input of sight, smell, or sound is
not an unmixed blessing, for it may lead to hallucinations of
sight, smell, or sound—phantopsia, phantosmia, or phantacu-
sis, to use the old but useful terms. And just as 10 to 20 percent
of those who lose their sight get Charles Bonnet syndrome, a
similar percentage of those who lose their sense of smell experience the olfactory equivalent. In some cases these phantom
smells follow sinus infections or head injuries, but occasionally they are associated with migraine, epilepsy, parkinsonism, PTSD, or other conditions.[2]

In CBS, if there is some remaining vision, there may also be

---

2. Among these other conditions is infection with the herpes simplex
virus, which can attack nerves (including sometimes the olfactory
nerves), both impairing and stimulating them. The virus can remain
dormant for long periods, sequestered in nerve ganglia, and suddenly
reemerge at intervals of months or years. One man, a microbiologist,
wrote to me: "In the summer of 2006, I began to 'smell things,' a faint
pervasive odor that I could not identify (my best guess was . . . wet cardboard)." Prior to this, he said, "I had a highly sensitive nose, and was able
to identify my laboratory cultures by smell alone, or subtle differences in
organic solvents, or faint perfumes."
He soon developed a constant hallucination of the smell of rotting
fish, which faded only after a year had passed, along with most of his
"olfactory acuity and the subtlety of most foods." He wrote:

> Certain odors are completely gone—feces(!), baking bread, or cook-
> ies, roasting turkey, garbage, roses, the fresh soil smell of *Strepto-
> myces* . . . all gone. I miss the smells of Thanksgiving, but not the
> smell of public toilets.

The dysosmia and phantosmia were due to a reemergence of the
herpes simplex 2 which he had contracted many years before, and he
is intrigued that these are always preceded by hallucinatory smells. He
writes, "I smell the onset of herpes reactivation. A day or two prior to the
onset of a neuritis episode, I again have olfactory hallucinations of the
last strong smell I noticed. [This smell] persists during the neuritis and
fades as the neuritis fades. . . . The strength of the hallucinations is correlated with the severity of the generalized neuritis."

perceptual distortions of all sorts. Similarly, those who have lost much but not all of their sense of smell tend to suffer from distortions of smell, often of an unpleasant sort (a condition called parosmia or dysosmia).

Mary B., a Canadian woman, acquired dysosmia two months after an operation performed under general anesthesia. Eight years later, she sent me a detailed account of her experiences, entitled "A Phantom in My Brain." She wrote:

It happened fast. In September 1999 I felt great. I'd had a hysterectomy in the summer, but I was already back to daily Pilates and ballet classes, feeling fit and full of vigour. Four months later I was still fit and vigorous, but I was locked in an invisible prison by a disorder no one could see, that no one seemed to know anything about, that I couldn't even find a name for.

The changes were gradual at first. In September tomatoes and oranges started tasting metallic and a bit rotten, and cottage cheese tasted like sour milk. I tried different brands; they were all bad.

During October, lettuce began to smell and taste of turpentine, and spinach, apples, carrots and cauliflower tasted slightly rotten. Fish and meat, especially chicken, smelt as if they'd been rotting for a week. My partner couldn't detect the off tastes at all. Was I developing some sort of food allergy? . . .

Soon the exhaust fans of restaurant kitchens started smelling weirdly unpleasant. Bread tasted rancid; chocolate, like machine oil. The only meat or fish I could eat was smoked salmon. I started having it three times a week. In early December we ate out with friends. I had to choose carefully, but I enjoyed the meal, except that the mineral water smelt like bleach. But the others were drinking it happily, and I decided that my glass hadn't been rinsed properly. Smells and tastes got dramatically worse in the next week. Traffic smelt so bad that I had to force myself to go out; I made long detours to go to my Pilates and ballet classes by pedestrian-only routes. Wine smelt revolting; so did anybody who was wearing scent. The smell of Ian's morning coffee had been getting worse, but between one day and the next it turned into a lurid, intolerable stench that permeated the house and lingered for hours. He started having coffee at work.

Ms. B. kept careful notes, hoping to find, if not an explanation, at least some pattern to the distortions, but she could find none. "There was no rhyme or reason to it," she wrote. "How could lemons taste okay but not oranges; garlic, but not onions?"

With complete anosmia, rather than exaggerations or distortions of perceived smells, there can only be hallucinations of smell. These too can be very various, and sometimes difficult to define or describe. This was brought out by Heather A.:

> The hallucinations generally cannot be described by one smell descriptor (except one night I smelled dill pickles for most of an evening). I can kind of describe them as an amalgam of other smells (metallic-y roll-on deodorant; dense acrid-sweet cake; melted plastic in a three-day-old garbage pile). I have been able to have fun with it in this way, make an art of naming/describing them. In the beginning, I would go through phases

where I would access one at a time for a couple of weeks, multiple times a day. After a few months, the family of smells I had gone through had diversified, and now I can reference several different ones in a day. Sometimes a new one will pop up and I may not smell it again. The experience of them varies. Sometimes they will come up strong, like something stuck right under my nose, and dissipate quickly; sometimes one will be subtle and linger, at times barely noticeable.

Some people hallucinate a particular smell, which may be influenced by context or suggestion. Laura H., who lost most of her sense of smell after a craniotomy, wrote to me that she would occasionally have a brief burst of smells that were plausible, though not always entirely accurate from what she remembered sensing before her loss. Sometimes they were not really there at all:

> Our kitchen was being revamped, and the electrics blew one evening. My husband assured me that all was safe but I was very worried about a possible electrical fire that might start. . . . I woke up in the middle of the night and had to get up to check the kitchen because I thought I could smell electrical burning. . . . I checked everywhere I could see in the kitchen, hall, cupboards, but could see nothing burning. . . . I then started to think the smell could be coming from behind a wall or somewhere I couldn't see.

She woke her husband; he could smell nothing, but she could still smell the smoke strongly. "I was shocked," she said, "by how strongly I could smell something that wasn't there."

Others may be haunted by a single constant smell of such

complexity that it seems to conglomerate almost all the bad smells in the world. Bonnie Blodgett, in her book *Remembering Smell*, describes the phantosmic world she was plunged into following a sinus infection and the use of a potent nasal spray. She was driving along a state highway when she first detected a "weird" smell. She checked her shoes at a gas station, found them clean, then wondered if there was something amiss with the heater fan in the car—a dead bird perhaps? The smell pursued her, waxing and waning in intensity but never absent. She explored a dozen possible external causes and was finally, reluctantly, forced to the realization that the smell was in her head—in a neurological, not a psychiatric, sense. She described the smell as resembling "shit, puke, burning flesh and rotten eggs. Not to mention smoke, chemicals, urine and mold. My brain had truly outdone itself." (Hallucination of particularly vile smells is called cacosmia.)

While humans can detect and identify perhaps ten thousand distinct smells, the number of possible smells is far greater, for there are more than five hundred different odorant receptor sites in the nasal mucosa, and stimulation of these (or their cerebral representations) may be combined in trillions of ways. Some hallucinated smells may be impossible to describe because they are different from anything ever experienced in the real world, and evoke no memories or associations. New, unprecedented experiences can be a hallmark of hallucinations, for when the brain is released from the constraints of reality, it can generate any sound, image, or smell in its repertoire, sometimes in complex and "impossible" combinations.

# 4

# Hearing Things

In 1973 the journal *Science* published an article that caused an immediate furor. It was entitled "On Being Sane in Insane Places," and it described how, as an experiment, eight "pseudopatients" with no history of mental illness presented themselves at a variety of hospitals across the United States. Their single complaint was that they "heard voices." They told hospital staff that they could not really make out what the voices said but that they heard the words "empty," "hollow," and "thud." Apart from this fabrication, they behaved normally and recounted their own (normal) past experiences and medical histories. Nonetheless, all of them were diagnosed as schizophrenic (except one, who was diagnosed with "manic-depressive psychosis"), hospitalized for up to two months, and prescribed antipsychotic medications (which they did not swallow). Once admitted to the mental wards, they continued to speak and behave normally; they reported to the medical staff that their hallucinated voices had disappeared and that they felt fine. They even kept notes on their experiment, quite openly (this was registered in the nursing notes for one pseudopatient as "writing behavior"), but none of the pseudopatients were identified as such by the staff.[1] This experiment, designed by David Rosenhan, a Stanford psychologist (and himself a pseudopatient), emphasized, among other things, that the single symptom of "hearing voices" could suffice for an immediate, categorical diagnosis of schizophrenia even in the absence of any other symptoms or abnormalities of behavior. Psychiatry, and society in general, had been subverted by the almost axiomatic belief that "hearing voices" spelled madness and never occurred except in the context of severe mental disturbance.

This belief is a fairly recent one, as the careful and humane reservations of early researchers on schizophrenia made clear. But by the 1970s, antipsychotic drugs and tranquilizers had begun to replace other treatments, and careful history taking, looking at the whole life of the patient, had largely been replaced by the use of DSM criteria to make snap diagnoses.

Eugen Bleuler, who directed the huge Burghölzli asylum near Zurich from 1898 to 1927, paid close and sympathetic attention to the many hundreds of schizophrenic people under his care. He recognized that the "voices" his patients heard, however outlandish they might seem, were closely associated with their mental states and delusions. The voices, he wrote, embodied "all their strivings and fears . . . their entire trans-

---

1. The real patients, however, were more observant. "You're not crazy," said one. "You're a journalist or a professor."

formed relationship to the external world . . . above all . . . [to] the pathological or hostile powers" that beset them. He described these in vivid detail in his great 1911 monograph, *Dementia Praecox; or, The Group of Schizophrenias:*

> The voices not only speak to the patient, but they pass electricity through the body, beat him, paralyse him, take his thoughts away. They are often hypostasized as people, or in other very bizarre ways. For example, a patient claims that a "voice" is perched above each of his ears. One voice is a little larger than the other but both are about the size of a walnut, and they consist of nothing but a large ugly mouth.
>
> Threats or curses form the main and most common content of the "voices." Day and night they come from everywhere, from the walls, from above and below, from the cellar and the roof, from heaven and from hell, from near and far. . . . When the patient is eating, he hears a voice saying, "Each mouthful is stolen." If he drops something, he hears, "If only your foot had been chopped off."
>
> The voices are often very contradictory. At one time they may be against the patient . . . then they may contradict themselves. . . . The roles of pro and con are often taken over by voices of different people. . . . The voice of a daughter tells a patient: "He is going to be burned alive," while his mother's voice says, "He will not be burned." Besides their persecutors the patients often hear the voice of some protector.
>
> The voices are often localized in the body. . . . A polyp may be the occasion for localizing the voices in the nose. An intestinal disturbance brings them into connection with the abdomen. . . . In cases of sexual complexes, the penis, the urine in the bladder, or the nose utter obscene words. . . . A really

or imaginarily gravid patient will hear her child or children speaking inside her womb. . . .

Inanimate objects may speak. The lemonade speaks, the patient's name is heard to be coming from a glass of milk. The furniture speaks to him.

Bleuler wrote, "Almost every schizophrenic who is hospitalized hears 'voices.' " But he emphasized that the reverse did not hold—that hearing voices did not necessarily denote schizophrenia. In the popular imagination, though, hallucinatory voices are almost synonymous with schizophrenia—a great misconception, for most people who do hear voices are not schizophrenic.

M any people report hearing voices which are not particularly directed at them, as Nancy C. wrote:

> I hallucinate conversations on a regular basis, often as I am falling asleep at night. It seems to me that these conversations are real and are actually taking place between real people, at the very time I'm hearing them, but are occurring somewhere else. I hear couples arguing, all kinds of things. They are not voices I can identify, they are not people I know. I feel like I'm a radio, tuned into someone else's world. (Though always an American-English-speaking world.) I can't think of any way to regard these experiences except as hallucinations. I am never a participant; I am never addressed. I am just listening in.

"Hallucinations in the sane" were well recognized in the nineteenth century, and with the rise of neurology, people sought to understand more clearly what caused them. In England in the 1880s, the Society for Psychical Research was founded to collect and investigate reports of apparitions or hallucinations, especially those of the bereaved, and many eminent scientists—physicists as well as physiologists and psychologists—joined the society (William James was active in the American branch). Telepathy, clairvoyance, communication with the dead, and the nature of a spirit world became the subjects of systematic investigation.

These early researchers found that hallucinations were not uncommon in the general population. Their 1894 "International Census of Waking Hallucinations in the Sane" examined the occurrence and nature of hallucinations experienced by normal people in normal circumstances (they took care to exclude anyone with obvious medical or psychiatric problems). Seventeen thousand people were sent a single question:

Have you ever, when believing yourself to be completely awake, had a vivid impression of seeing or being touched by a living being or inanimate object, or of hearing a voice, which impression, as far as you could discover, was not due to an external physical cause?

More than 10 percent responded in the affirmative, and of those, more than a third heard voices. As John Watkins noted in his book *Hearing Voices*, hallucinated voices "having some kind of religious or supernatural content represented a small but significant minority of these reports." Most of the hallucinations, however, were of a more quotidian character.

Perhaps the commonest auditory hallucination is hearing one's own name spoken—either by a familiar voice or an anon-

ymous one. Freud, writing in *The Psychopathology of Everyday Life*, remarked on this:

> During the days when I was living alone in a foreign city—I
> was a young man at the time—I quite often heard my name
> suddenly called by an unmistakable and beloved voice; I then
> noted down the exact moment of the hallucination and made
> anxious enquiries of those at home about what had happened at
> that time. Nothing had happened.[2]

The voices that are sometimes heard by people with schizophrenia tend to be accusing, threatening, jeering, or persecuting. By contrast, the voices hallucinated by the "normal" are
often quite unremarkable, as Daniel Smith brings out in his
book *Muses, Madmen, and Prophets: Hearing Voices and the
Borders of Sanity*. Smith's own father and grandfather heard
such voices, and they had very different reactions. His father
started hearing voices at the age of thirteen, Smith writes:

> These voices weren't elaborate, and they weren't disturbing
> in content. They issued simple commands. They instructed
> him, for instance, to move a glass from one side of the table to
> another or to use a particular subway turnstile. Yet in listening
> to them and obeying them his interior life became, by all
> reports, unendurable.

Smith's grandfather, by contrast, was nonchalant, even playful, in regard to his hallucinatory voices. He described how he

---

2. Freud was not unsympathetic to the notion of telepathy; his "Psy-
choanalysis and Telepathy" was written in 1921, though published only
posthumously.

tried to use them in betting at the racetrack. ("It didn't work, my mind was clouded with voices telling me that this horse could win or maybe this one is ready to win.") It was much more successful when he played cards with his friends. Neither the grandfather nor the father had strong supernatural inclinations; nor did they have any significant mental illness. They just heard unremarkable voices concerned with everyday things—as do millions of others.

Smith's father and grandfather rarely spoke of their voices. They listened to them in secrecy and silence, perhaps feeling that admitting to hearing voices would be seen as an indication of madness or at least serious psychiatric turmoil. Yet many recent studies confirm that it is not that uncommon to hear voices and that the majority of those who do are not schizophrenic; they are more like Smith's father and grandfather.[3]

It is clear that attitudes to hearing voices are critically important. One can be tortured by voices, as Daniel Smith's father was, or accepting and easygoing, like his grandfather. Behind these personal attitudes are the attitudes of society, attitudes which have differed profoundly in different times and places.

Hearing voices occurs in every culture and has often been accorded great importance—the gods of Greek myth

---

3. Recently, a number of people who hear voices have organized networks in various countries asserting their "right" to hear voices, to have them respected and not dismissed as trivial or pathological. This movement and its significance are discussed by Ivan Leudar and Philip Thomas in their book *Voices of Reason, Voices of Madness* and by Sandra Escher and Marius Romme in their 2012 review of the subject.

often spoke to mortals, and the gods of the great monotheistic traditions, too. Voices have been significant in this regard, perhaps more so than visions, for voices, language, can convey an explicit message or command as images alone cannot.

Until the eighteenth century, voices—like visions—were ascribed to supernatural agencies: gods or demons, angels or djinns. No doubt there was sometimes an overlap between such voices and those of psychosis or hysteria, but for the most part, voices were not regarded as pathological; if they stayed inconspicuous and private, they were simply accepted as part of human nature, part of the way it was with some people.

Around the middle of the eighteenth century, a new secular philosophy started to gain ground with the philosophers and scientists of the Enlightenment, and hallucinatory visions and voices came to be seen as having a physiological basis in the overactivity of certain centers in the brain.

But the romantic idea of "inspiration" still held, too—the artist, especially the writer, was seen or saw himself as the transcriber, the amanuensis, of a Voice, and sometimes had to wait years (as Rilke did) for the Voice to speak.[4]

Talking to oneself is basic to human beings, for we are a linguistic species; the great Russian psychologist Lev Vygotsky thought that "inner speech" was a prerequisite of all voluntary activity. I talk to myself, as many of us do, for much of the day—admonishing myself ("You fool! Where did

---

4. Judith Weissman, in her book *Of Two Minds: Poets Who Hear Voices*, presents strong evidence, drawn especially from what poets themselves have said, that many of them, from Homer to Yeats, have been inspired by true auditory vocal hallucinations, not just metaphorical voices.

you leave your glasses?"), encouraging myself ("You can do it!"), complaining ("Why is that car in my lane?"), and, more rarely, congratulating myself ("It's done!"). Those voices are not externalized; I would never mistake them for the voice of God, or anyone else.

But when I was in danger once, trying to descend a mountain with a badly injured leg, I heard an inner voice that was wholly unlike my normal babble of inner speech. I had a great struggle crossing a stream with a buckled and dislocating knee. The effort left me stunned, motionless for a couple of minutes, and then a delicious languor came over me, and I thought to myself, Why not rest here? A nap maybe? This was immediately countered by a strong, clear, commanding voice, which said, "You can't rest here—you can't rest anywhere. You've got to go on. Find a pace you can keep up and go on steadily." This good voice, this Life voice, braced and resolved me. I stopped trembling and did not falter again.

Joe Simpson, climbing in the Andes, also had a catastrophic accident, falling off an ice ledge and ending up in a deep crevasse with a broken leg. He struggled to survive, as he recounted in *Touching the Void*—and a voice was crucial in encouraging and directing him:

There was silence, and snow, and a clear sky empty of life, and me, sitting there, taking it all in, accepting what I must try to achieve. There were no dark forces acting against me. A voice in my head told me that this was true, cutting through the jumble in my mind with its coldly rational sound.

It was as if there were two minds within me arguing the toss. The *voice* was clean and sharp and commanding. It was always right, and I listened to it when it spoke and acted on its decisions. The other mind rambled out a disconnected series of images, and memories and hopes, which I attended to in a daydream state as I set about obeying the orders of the *voice*. I had to get to the glacier. . . . The *voice* told me exactly how to go about it, and I obeyed while my other mind jumped abstractly from one idea to another. . . . The *voice,* and the watch, urged me into motion whenever the heat from the glacier halted me in a drowsy exhausted daze. It was three o'clock—only three and a half hours of daylight left. I kept moving but soon realized that I was making ponderously slow headway. It didn't seem to concern me that I was moving like a snail. So long as I obeyed the *voice*, then I would be all right.

Such voices may occur with anyone in situations of extreme threat or danger. Freud heard voices on two such occasions, as he mentioned in his book *On Aphasia*:

I remember having twice been in danger of my life, and each time the awareness of the danger occurred to me quite suddenly. On both occasions I felt "this was the end," and while otherwise my inner language proceeded with only indistinct sound images and slight lip movements, in these situations of danger I heard the words as if somebody was shouting them into my ear, and at the same time I saw them as if they were printed on a piece of paper floating in the air.

The threat to life may also come from within, and although we cannot know how many attempts at suicide have been prevented by a voice, I suspect this is not uncommon. My friend Liz, following the collapse of a love affair, found herself heartbroken and despondent. About to swallow a handful of sleeping tablets and wash them down with a tumbler of whiskey, she was startled to hear a voice say, "No. You don't want to do that," and then "Remember that what you are feeling now you will not be feeling later." The voice seemed to come from the outside; it was a man's voice, though whose she did not know. She said, faintly, "Who said that?" There was no answer, but a "granular" figure (as she put it) materialized in the chair opposite her—a young man in eighteenth-century dress who glimmered for a few seconds and then disappeared. A feeling of immense relief and joy came over her. Although Liz knew that the voice must have come from the deepest part of herself, she speaks of it, playfully, as her "guardian angel."

Various explanations have been offered for why people hear voices, and different ones may apply in different circumstances. It seems likely, for example, that the predominantly hostile or persecuting voices of psychosis have a very different basis from the hearing of one's own name called in an empty house; and that this again is different in origin from the voices which come in emergencies or desperate situations.

Auditory hallucinations may be associated with abnormal activation of the primary auditory cortex; this is a subject which needs much more investigation not only in those with psychosis but in the population at large—the vast majority of studies so far have examined only auditory hallucinations in psychiatric patients.

Some researchers have proposed that auditory hallucinations result from a failure to recognize internally generated speech as one's own (or perhaps it stems from a cross-activation with the auditory areas so that what most of us experience as our own thoughts becomes "voiced").

Perhaps there is some sort of physiological barrier or inhibition that normally prevents most of us from "hearing" such inner voices as external. Perhaps that barrier is somehow breached or undeveloped in those who do hear constant voices. Perhaps, however, one should invert the question—and ask why most of us do not hear voices. Julian Jaynes, in his influential 1976 book, *The Origin of Consciousness in the Breakdown of the Bicameral Mind*, speculated that, not so long ago, all humans heard voices—generated internally, from the right hemisphere of the brain, but perceived (by the left hemisphere) as if external, and taken as direct communications from the gods. Sometime around 1000 B.C., Jaynes proposed, with the rise of modern consciousness, the voices became internalized and recognized as our own.[5]

Others have proposed that auditory hallucinations may come from an abnormal attention to the subvocal stream which accompanies verbal thinking. It is clear that "hearing voices" and "auditory hallucinations" are terms that cover a variety of different phenomena.

While voices carry meaning—whether this is trivial or portentous—some auditory hallucinations consist of

---

5. Jaynes thought that there might be a reversion to "bicamerality" in schizophrenia and some other conditions. Some psychiatrists (such as Nasrallah, 1985) favor this idea or, at the least, the idea that the hallucinatory voices in schizophrenia emanate from the right side of the brain but are not recognized as one's own, and are thus perceived as alien.

little more than odd noises. Probably the most common of these are classified as tinnitus, an almost nonstop hissing or ringing sound that often goes with hearing loss, and may be intolerably loud at times.

Hearing noises—hummings, mutterings, twitterings, rappings, rustlings, ringings, muffled voices—is commonly associated with hearing problems, and this may be aggravated by many factors, including delirium, dementia, toxins, or stress. When medical residents, for example, are on call for long periods, sleep deprivation may produce a variety of hallucinations involving any sensory modality. One young neurologist wrote to me that after being on call for more than thirty hours, he would hear the hospital's telemetry and ventilator alarms, and sometimes after arriving home he kept hallucinating the phone ringing.[6]

Although musical phrases or songs may be heard along with voices or other noises, a great many people "hear" only music or musical phrases. Musical hallucinations may arise from a stroke, a tumor, an aneurysm, an infectious disease, a neurodegenerative process, or toxic or metabolic disturbances. Hallucinations in such situations usually disappear as soon as the provocative cause is treated or subsides.[7]

6. Sarah Lipman has noted, in her blog (www.reallysarahsyndication .com), the phenomenon of "phantom rings" as people imagine or hallucinate the ringing of their cell phones. She links this to a state of vigilance, expectation, or anxiety, as when she thinks she may hear a knock at the door or her baby crying. "Part of my consciousness," she wrote to me, "is straining to monitor for the sound. It seems to me that it is this hyper-alert state that generates the phantom sounds."

7. There may be paroxysmal musical hallucinations during temporal

Sometimes it is difficult to pinpoint a particular cause for musical hallucinations, but in the predominantly geriatric population I work with, by far the commonest cause of musical hallucination is hearing loss or deafness—and here the hallucinations may be stubbornly persistent, even if the hearing is improved by hearing aids or cochlear implants. Diane G. wrote to me:

I have had tinnitus as far back as I can remember. It is present almost 24/7 and is very high pitched. It sounds exactly like how cicadas sound when they come in droves back on Long Island in the summer. Sometime in the last year [I also became aware of] the music playing in my head. I kept hearing Bing Crosby, friends and orchestra singing "White Christmas" over and over. I thought it was coming from a radio playing in another room until I eliminated all possibilities of outside input. It went on for days, and I quickly discovered that I could not turn it off or vary the volume. But I could vary the lyrics, speed and harmonies with practice. Since that time I get the music almost daily, usually toward evenings and at times so loud that it interferes with my hearing conversations. The music is always melodies that I am familiar with such as hymns, favorites from years of piano playing and songs from early memories. They always have the lyrics. . . .

To add to this cacophony, I now have started hearing a third level of sound at the same time that sounds like someone is

---

lobe seizures. But in such cases, the musical hallucinations have a fixed and invariable format; they appear along with other symptoms (perhaps visual or olfactory hallucinations or a sense of déjà vu) and at no other time. If the seizures can be controlled medically or surgically, the epileptic music will cease.

listening to talk radio or TV in another room. I get a constant running of voices, male and female, complete with realistic pauses, inflections and increases and decreases in volume. I just can't understand their words.

Diane has had progressive hearing loss since childhood, and she is unusual in that she has hallucinations of both music and conversation.[8]

There is a wide range in the quality of individual musical hallucinations—sometimes they are soft, sometimes disturbingly loud; sometimes simple, sometimes complex—but there are certain characteristics common to all of them. First and foremost, they are perceptual in quality and seem to emanate from an external source; in this way they are distinct from imagery (even "earworms," the often annoying, repetitious musical imagery that most of us are prone to from time to time). People with musical hallucinations will often search for an external cause—a radio, a neighbor's television, a band in the street—and only when they fail to find any such external source do they realize that the source must be in themselves. Thus they may liken it to a tape recorder or an iPod

8. Most people who get musical hallucinations are elderly and somewhat deaf; it is not unusual for them to be treated as if demented, psychotic, or imbecilic. Jean G. was hospitalized after she had an apparent heart attack, and a few days later, she began "hearing a male choir in the distance as if it were coming through the woods." (Several years later, when she wrote to me, she still heard this, especially in times of stress or when she was extremely tired.) But, she said, "I quickly stopped talking about this type of music when faced with a nurse asking me, 'Do you know your name? Do you know what day this is?' I responded back, 'Yes, I know what day this is—it is the day I am going home.' "

in the brain, something mechanical and autonomous, not a controllable, integral part of the self.

That there should be something like this in one's head arouses bewilderment and, not infrequently, fear—fear that one is going mad or that the phantom music may be a sign of a tumor, a stroke, or a dementia. Such fears often inhibit people from acknowledging that they have hallucinations; perhaps for this reason musical hallucinations have long been considered rare—but it is now realized that this is far from the case.[9]

Musical hallucinations can intrude upon and even overwhelm perception; like tinnitus, they can be so loud as to make it impossible to hear someone speak (imagery almost never competes with perception in this way).

Musical hallucinations often appear suddenly, with no apparent trigger. Frequently, however, they follow a tinnitus or an external noise (like the drone of a plane engine or a lawn mower), the hearing of real music, or anything suggestive of a particular piece or style of music. Sometimes they are triggered by external associations, as with one patient of mine who, whenever she passed a French bakery, would hear the song "Alouette, gentille alouette."

Some people have musical hallucinations virtually nonstop, while others have them only intermittently. The hallucinated music is usually familiar (though not always liked; thus one of my patients hallucinated Nazi marching songs from his youth, which terrified him). It may be vocal or instrumental, classical or popular, but it is most often music heard in the patient's early years. Occasionally, patients may hear "meaningless

---

9. I have written at much greater length about musical hallucinations (as well as intrusive musical imagery, or "earworms") in my book *Musicophilia.*

phrases and patterns," as one of my correspondents, a gifted musician, put it.

Hallucinated music can be very detailed, so that every note in a piece, every instrument in an orchestra, is distinctly heard. Such detail and accuracy is often astonishing to the hallucinator, who may be scarcely able, normally, to hold a simple tune in his head, let alone an elaborate choral or instrumental composition. (Perhaps there is an analogy here to the extreme clarity and unusual detail which characterize many visual hallucinations.) Often a single theme, perhaps only a few bars, is hallucinated again and again, like a skipping record. One patient of mine heard part of "O Come, All Ye Faithful" nineteen and a half times in ten minutes (her husband timed this) and was tormented by never hearing the entire hymn. Hallucinatory music can wax slowly in intensity and then slowly wane, but it may also come on suddenly full blast in mid-bar and then stop with equal suddenness (like a switch turned on and off, patients often comment). Some patients may sing along with their musical hallucinations; others ignore them—it makes no difference. Musical hallucinations continue in their own way, irrespective of whether one attends to them or not. And they can continue, pursuing their own course, even if one is listening to or playing something else. Thus Gordon B., a violinist, sometimes hallucinated a piece of music while he was actually performing an entirely different piece at a concert.

Musical hallucinations tend to spread. A familiar tune, an old song, may start the process; this is likely to be joined, over a period of days or weeks, by another song, and then another, until a whole repertoire of hallucinatory music has been built up. And this repertoire itself tends to change—one tune will drop out, and another will replace it. One cannot voluntarily start or stop the hallucinations, though some people may be able, on occasion, to replace one piece of hallucinated music with another. Thus one man who said he had "an intracranial jukebox" found that he could switch at will from one "record" to another, provided there was some similarity of style or rhythm, though he could not turn on or turn off the "jukebox" as a whole.

Prolonged silence or auditory monotony may also cause auditory hallucinations; I have had patients report experiencing these while on meditation retreats or on a long sea voyage. Jessica K., a young woman with no hearing loss, wrote to me that her hallucinations come with auditory monotony:

> In the presence of white noise such as running water or a central air conditioning system, I frequently hear music or voices. I hear it distinctly (and in the early days, often went searching for the radio that must have been left on in another room), but in the instance of music with lyrics or voices (which always sound like a talk radio program or something, not real conversation) I never hear it well enough to distinguish the words. I never hear these things unless they are "embedded," so to speak, in white noise, and only if there are not other competing sounds.

Musical hallucinations seem to be less common in children, but one boy I have seen, Michael, has had them since the age of five or six. His music is nonstop and overwhelming, and it often prevents him focusing on anything else. Much more often, musical hallucinations are acquired at a later age—unlike hearing voices, which seems, in those who have it, to begin in early childhood and to last a lifetime.

Some people with persistent musical hallucinations find them tormenting, but most people accommodate and learn to live with the music forced on them, and a few even come to enjoy their internal music and may feel it as an enrichment of life. Ivy L., a lively and articulate eighty-five-year-old, has had some visual hallucinations related to her macular degeneration, and some musical and auditory hallucinations stemming from her hearing impairment. Mrs. L. wrote to me:

In 2008 my doctor prescribed paroxetine for what she called depression and I called sadness. I had moved from St. Louis to Massachusetts after my husband died. A week after starting paroxetine, while watching the Olympics, I was surprised to hear languid music with the men's swim races. When I turned off the TV, the music continued and has been present virtually every waking minute since.

When the music began, a doctor gave me Zyprexa as a possible aid. That brought a visual hallucination of a murky, bubbling brown ceiling at night. A second prescription gave me hallucinations of lovely, transparent tropical plants growing in my bathroom. So I quit taking these prescriptions and the visual hallucinations ceased. The music continued.

I do not simply "recall" these songs. The music playing in the house is as loud and clear as any CD or concert. The volume increases in a large space such as a supermarket. The music has no singers or words. I have never heard "voices" but once heard my name called urgently, while I was dozing.

There was a short time when I "heard" doorbells, phones, and alarm clocks ring although none were ringing. I no longer experience these. In addition to music, at times I hear katydids, sparrows, or the sound of a large truck idling at my right side.

During all these experiences, I am fully aware that they are not real. I continue to function, managing my accounts and finances, moving my residence, taking care of my household. I speak coherently while experiencing these aural and visual disturbances. My memory is quite accurate, except for the occasional misplaced paper.

I can "enter" a melody I think of or have one triggered by a phrase, but I cannot stop the aural hallucinations. So I cannot stop the "piano" in the coat closet, the "clarinet" in the living room ceiling, the endless "God Bless America"s, or waking up to "Good Night, Irene." But I manage.

PET and fMRI scanning have shown that musical hallucination, like actual musical perception, is associated with the activation of an extensive network involving many areas of the brain—auditory areas, motor cortex, visual areas, basal ganglia, cerebellum, hippocampi, and amygdala. (Music calls upon many more areas of the brain than any other activity—one reason why music therapy is useful for such a wide variety of conditions.) This musical network can be stimulated directly, on occasion, as by a focal epilepsy, a fever, or delirium, but what seems to occur in most cases of musical hallucinations is a release of activity in the musical network when normally operative inhibitions or constraints are weakened. The commonest cause of such a release is auditory deprivation or deafness. In this way, the musical hallucinations of the elderly deaf are analogous to the visual hallucinations of Charles Bonnet syndrome.

But although the musical hallucinations of deafness and the visual hallucinations of CBS may be akin physiologically, they have great differences phenomenologically, and these reflect the very different nature of our visual worlds and our musical worlds—differences evident in the ways we perceive, recollect, or imagine them. We are not given an already made, pre-assembled visual world; we have to construct our own visual world as best we can. This construction entails analysis and synthesis at many functional levels in the brain, starting with perception of lines and angles and orientation in the occipital cortex. At higher levels, in the inferotemporal cortex, the "elements" of visual perception are of a more complex sort, appropriate for the analysis and recognition of natural scenes, objects, animal and plant forms, letters, and faces. Complex visual hallucinations entail the putting together of such elements, an act of assemblage, and these assemblages are continually permuted, disassembled, and reassembled.

Musical hallucinations are quite different. With music, although there are separate functional systems for perceiving pitch, timbre, rhythm, etc., the musical networks of the brain work together, and pieces cannot be significantly altered in melodic contour or tempo or rhythm without losing their musical identity. We apprehend a piece of music as a whole. Whatever the initial processes of musical perception and memory may be, once a piece of music is known, it is retained not as an assemblage of individual elements but as a completed procedure or performance; music is *performed* by the mind/brain whenever it is recollected; and this is also so when it erupts spontaneously, whether as an earworm or as a hallucination.

5

# The Illusions
# of Parkinsonism

James Parkinson, in his famous 1817 *Essay on the Shaking Palsy*, portrayed the disease that now bears his name as one that affected movement and posture, while leaving the senses and the intellect unimpaired. And in the century and a half that followed, there was virtually no mention of perceptual disorders or hallucinations in patients with Parkinson's disease. By the late 1980s, though, physicians had begun to realize (and only in response to careful inquiry, for patients are often reluctant to admit it) that perhaps a third or more of those being treated for Parkinson's experienced hallucinations, as Gilles Fénelon and others reported. By this time, virtually everyone diagnosed with Parkinson's was medicated with L-dopa or other drugs that enhance the neurotransmitter dopamine in the brain.

My own experience with parkinsonism as a young doctor was predominantly with the patients I described in *Awakenings*, who did not have ordinary Parkinson's disease but a much more complex syndrome. They were survivors of the encephalitis lethargica epidemic that followed the First World War, and they had come down, sometimes decades later, with postencephalitic syndromes including not only a very severe form of parkinsonism but often a host of other disorders, especially sleep and arousal disorders. These postencephalitic patients were far more sensitive to the effects of L-dopa than patients with ordinary Parkinson's disease. Many of them, once they were started on L-dopa, began to have excessively vivid dreams or nightmares; often this would be the first apparent effect of the medication. Several of them became prone to visual illusions or hallucinations, too.

When Leonard L. was started on L-dopa, he began to see faces on the blank screen of his television set, and a picture of an old western town that hung in his room would come to life as he looked at it, with people emerging from its saloons and cowboys galloping through the streets.

Martha N., another postencephalitic patient, would "sew" with hallucinatory needles and thread. "See what a lovely coverlet I have stitched for you today!" she said on one occasion. "See the pretty dragons, the unicorn in his paddock." She traced their invisible outlines in the air. "Here, take it," she said, and placed the ghostly thing in my hands.

With Gertie C., the hallucinations (precipitated by the addition of amantadine to her L-dopa) were less benign. Within three hours of receiving the first dose, she became intensely excited and deliriously hallucinated. She would cry out, "Cars bearing down on me, they're crowding me!" She also saw faces "like masks popping in and out." Occasionally she would smile rapturously and exclaim, "Look what a beautiful tree, so beautiful," and tears of pleasure would fill her eyes.

In contrast to these postencephalitic patients, people with ordinary Parkinson's disease do not usually experience visual hallucinations until they have been on medication for many months or years. By the 1970s, I had several such patients who had started to get hallucinations, which were predominantly (though not exclusively) visual. Sometimes these began as webs and filigrees or other geometrical patterns; other patients experienced complex hallucinations, usually of animals and people, from the start. Such visions might seem quite real (one patient had a nasty fall while chasing a hallucinatory mouse), but the patients soon learned to distinguish them from reality and ignore them. At the time I could find almost nothing in the medical literature about such hallucinations, although it was sometimes said that L-dopa might make patients "psychotic." But by 1975, more than a quarter of my patients with ordinary Parkinson's disease, while otherwise doing well on L-dopa and dopamine agonists, had found themselves living with hallucinations.

Ed W., a designer, started to get visual hallucinations after he had been on L-dopa and dopamine agonists for several years. He realized that they were hallucinations and regarded them largely with curiosity and amusement; nevertheless, one of his physicians declared him "psychotic"—an upsetting misdiagnosis.

He often feels himself "on the verge" of hallucination, and he may be pushed over the threshold at night, or if he is tired or bored. When we had lunch one day, he was having all sorts of what he calls "illusions." My blue pullover, draped over a chair, became a fierce chimerical animal with an elephant-like head, long blue teeth, and a hint of wings. A bowl of noodles on the table became "a human brain" (though this did not affect his appetite for them). He saw "letters, like teletype" on my lips; they formed "words"—words he could not read. They did not coincide with the words I was speaking. He says that such illusions are "made up" on the spot, instantaneously and without conscious volition. He cannot control or stop them, short of closing his eyes. They are sometimes friendly, sometimes frightening. For the most part, he ignores them.

Sometimes he moves from "illusions" to frank hallucinations. One such was a hallucination of his cat, which had gone to the vet for a few days. Ed continued to "see" her at home, several times a day, emerging from the shadows at one end of the room. She would walk across the room, paying no attention to him, and then disappear into the shadows again. Ed realized at once that this was a hallucination, and had no desire to interact with it (though it aroused his curiosity and interest). When the real cat came back, the phantom cat disappeared.[1]

In addition to such isolated or occasional hallucinations, people with Parkinson's may develop elaborate and frighten-

---

1. My colleague Steven Frucht described to me a hallucination experienced by a patient of his, an intellectually intact woman who has been treated with medications for Parkinson's disease for more than fifteen years. Her hallucinations, however, started only a year ago. She also sees a cat—a grey cat with "beautiful" eyes which wears a serene, "beautiful expression" on its face and seems to be of a most friendly disposition. To her own surprise (for she has never liked cats), she enjoys visits from the grey cat and worries that "something may happen to him." Though she knows the cat is a hallucination, he seems very real to her: she can hear him coming, feel the warmth of his body, and touch him if she wishes. The first time the cat appeared, wanting to rub against her legs, she said, "Don't touch me, don't get too close." And since then the cat has kept a decorous distance. Occasionally, in the afternoon, the cat is joined by a large black dog. When Dr. Frucht asked her what happens when the cat sees the dog, she replied that the cat "looks away and is peaceful." She later remarked, "He is fulfilling his purpose in coming to visit me."

ing hallucinations, often of a paranoid sort. Such a psychosis took hold of Ed toward the end of 2011. He started to have hallucinations of people who entered his apartment, emerging from "a secret chamber" behind the kitchen. "They invade my privacy," Ed said. "They occupy my space. . . . They are very interested in me—they take notes, take photos, rifle through my papers." Sometimes they had sex—one of the intruders was a very beautiful woman, and sometimes three or four of them would occupy Ed's bed when he was not using it. These apparitions never appeared if he had real visitors or when he was listening to music or watching a favorite TV show; nor would they follow him when he left his apartment. He often regarded these persecutors as real and might say to his wife, "Take a cup of coffee to the man in my office." She always knew when he was hallucinating—he would stare fixedly at one point or follow an invisible presence with his eyes. Increasingly, he started to talk with them—or *at* them, for they never replied.

Ed's neurologist, on hearing this, advised him to have "a drug holiday," to stop all his anti-Parkinson's medications for two or three weeks, but this left Ed so incapacitated he could hardly move or speak. He then planned a gradual reduction of medication, and, two months later, on half his previous dose of L-dopa, Ed's hallucinations, his fears, and his psychosis have cleared completely.

In 2008, Tom C., an artist, came to my office for a consultation. He had been diagnosed with Parkinson's disease and put on medication about fifteen years earlier. Two years later, he started to experience "misperceptions," as he calls them (like the others, he avoids the term "hallucinations"). He is fond of dancing—he finds that this can unfreeze him, releasing him, for a while, from his parkinsonism. His first misperceptions occurred when he was in a nightclub; the skin of the other dancers, even their faces, seemed to be covered with tattoos. At first he thought the tattoos were real, but they started to glow and then to pulse and writhe; at that point he realized they must be hallucinatory. As an artist and a psychologist, he was intrigued by this experience—but frightened, too, that it might be the beginning of uncontrollable hallucinations of all sorts.

Once, while sitting at his desk, he was surprised to see a picture of the Taj Mahal on his computer monitor. As he gazed, the picture became richer in color, three-dimensional, vividly real. He heard a vague chanting, of the sort he thought might be associated with an Indian temple.

Another day, while he was lying on the floor, frozen by his parkinsonism, the reflections on a fluorescent ceiling lamp started to change into old photographs, mostly in black and white. These seemed to be photographs from earlier days, mostly of family, with some of strangers. "I had nothing else to do" in this immobilized state, he said, so he happily indulged this mild hallucinatory pleasure.

For Ed W. and Tom C., hallucinations usually remain on the "misperception" side, but Agnes R., a seventy-five-year-old lady who has had Parkinson's disease for twenty years, has had frank visual hallucinations for the last decade. She is, as she says, "an old hand" at hallucinations: "I see a whole array of things, which I enjoy—they are fascinating; they don't frighten me." In the clinic waiting room, she had seen "women—five of them—trying on fur coats." The size, color, solidity, and movement of these women looked perfectly natural; they seemed absolutely real. She knew that they were hallucinations only because they were out of context: no one would be trying on fur coats on a summer day in a doctor's office. In general, she is able to distinguish her hallucinations from reality, but there have been exceptions: on one occasion, seeing a furry black animal leap onto the dining table, she jumped. At other times, while walking, she has stopped abruptly to avoid bumping into a hallucinatory figure just in front of her.

Agnes most often sees apparitions from the windows of her twenty-second-floor apartment. From here, she has "seen" a skating rink on top of a (real) church, "people in tennis courts" on neighboring rooftops, and men working just outside her window. She does not recognize any of the people she sees, and they continue whatever they are doing without paying any attention to her. She watches these hallucinatory scenes with equanimity and sometimes with enjoyment. (Indeed, I got the impression that they help her pass the time—time which now seems to pass more slowly with her relative immobility and difficulties reading.) Her visions are not like dreams, she said; nor do they resemble fantasies. She has a great love for travel and for Egypt in particular, but she has never had "Egyptian" hallucinations or travel ones.

She sees no patterns to her hallucinations—they may come at any time of day, when she is busy with others or when she is alone. They seem to have nothing to do with current events, with her feelings, thoughts, or moods, or with the time of day she takes her medication. She cannot will them to come, or will them away. They superimpose themselves on what she is looking at and vanish, along with actual visual perception, when she closes her eyes.

Ed W. often describes a persistent feeling of a "presence"—something or someone he never actually sees—on his right. Professor R., while doing very well on L-dopa and other anti-parkinsonian drugs, also has "a companion" (as he calls it), just out of sight on his right. The sense of someone there is so strong that he sometimes wheels round to look, though there is never anyone to be seen. But his chief illusion is the transformation of print, words and sentences, into musical notation. The first time this happened was about two years ago. He was reading a book, turned away for a few seconds, and found, when he returned to the book, that the print had been replaced by music. This has occurred many times since, and may also be induced by staring at a page of print. Occasionally the dark border of his bath mat turns into staves and lines of music. There is always something—letters or lines—that is transformed into music; this may be why he regards these as "illusions," not hallucinations.

Professor R. is a very good musician; he started to play the piano when he was five and still plays for many hours a day. He is curious about his illusions, and he has done his best to transcribe or play the illusory music. (His best chance of "catching" this phantom music, he has found, is to set up a newspaper on the music stand and play it as soon as the newsprint turns into music.) But the "music" is scarcely playable, because it is always very highly ornamented, with innumerable crescendo and decrescendo markings, while the melody line is three or more octaves above middle C, and so may have half a dozen or more ledger lines above the treble staff.

Seeing music has been described to me by others (see

pages 13–15). Esther B., a composer and music teacher, wrote to me that, twelve years after being diagnosed with Parkinson's, she started to have "a rather peculiar visual phenomenon." She described this in detail:

> When I look at a surface—like a wall, or a floor, or a garment someone is wearing, or a curved surface like a tub or sink, or other surfaces too numerous to mention—I see a sort of collage of music scores superimposed upon the surface, especially with my peripheral vision. When I try to focus on any one image, it dims out or disappears elusively. These images of music scores come unbidden and are especially vivid after I've been working with any written music. The images always appear more or less horizontal, and if I tilt my head one way or the other, the horizontal images will tilt accordingly.

Howard H., a psychotherapist, began to notice tactile hallucinations soon after being diagnosed with Parkinson's, as he wrote:

> I would feel that the surfaces of various objects were covered by a film of fuzz, like peach fuzz, or the down in a pillow. It could also be described as cotton candy or spider webs. Sometimes the webs and fuzz can get very "lush," as when I reach down to pick up something that has fallen under my desk and my hand feels as if it has become submerged in a huge pile of this "stuff." But when I try to scoop up that pile, I see nothing and yet feel that I have a large amount of this "stuff" in my hands.

Is the use of L-dopa wholly responsible for such effects? Can L-dopa be regarded as a hallucinogenic drug? This seems unlikely in view of the fact that it is used in treating other conditions (such as dystonias) without provoking any hallucinations. Is there then something in the parkinsonian brain, or at least in some parkinsonian brains, which may predispose to visual hallucinations?[2]

Too often, parkinsonism is seen as no more than a movement disorder, but it may also involve a number of other aspects, including sleep disturbances of various sorts. People with Parkinson's may sleep poorly at night and often have chronic sleep deprivation. Their sleep may be marked by vivid and sometimes bizarre dreams, or nightmares in which they are awake but paralyzed, helpless to combat dream images being superimposed on their waking consciousness. All of these factors additionally predispose to hallucination.

In 1922 the French neurologist Jean Lhermitte described the sudden onset of visual hallucinations in an elderly patient—people in costume, children playing, animals around her (she would sometimes try to touch them). The patient had insomnia at night and drowsiness in the daytime, and her hallucinations tended to come at dusk.

Though this lady had dramatic visual hallucinations, she

---

2. Impairment of the sense of smell may appear early in Parkinson's disease and may perhaps predispose to smell hallucinations as well. But even in the absence of a noticeable impairment of smell, as Landis and Burkhard suggested in a 2008 paper, patients with incipient Parkinson's disease may have olfactory hallucinations before they develop motor symptoms.

had no visual impairments and no lesions in the visual cortex. But she had neurological signs suggesting an unusual pattern of damage in parts of the brain stem, the midbrain, and the pons. It was well known at this point that lesions in the visual pathway could cause hallucinations, but it was not clear how damage in the midbrain—not a visual area—could do so. Lhermitte thought such hallucinations might go with a derangement of the sleep-wake cycle, that they were essentially dreams or dream fragments invading daytime consciousness.

Five years later, the Belgian neurologist Ludo van Bogaert reported a somewhat similar case—his patient suddenly started seeing the heads of animals projected on the walls of his house at dusk. There were neurological signs similar to those of Lhermitte's patient, and van Bogaert also surmised midbrain damage. When his patient died, a year later, an autopsy revealed a large midbrain infarction involving (among other structures) the cerebral peduncles (hence his coinage of the term "peduncular hallucinations").

In Parkinson's disease, postencephalitic parkinsonism, and Lewy body disease, there is damage to the brain stem and associated structures, as there is in peduncular hallucinosis—though the damage occurs gradually and not abruptly, as with a stroke. In all of these degenerative diseases, however, there may be hallucinations, as well as sleep, movement, and cognitive disorders. But the hallucinations are markedly different from those of CBS; they are nearly always complex rather than elementary, often multisensory, and more apt to lead to delusions, which CBS alone rarely does. The hallucinations of brain-stem origin seem to be associated with abnormalities in the acetylcholine transmitter system—abnormalities that may be aggravated by giving the patient L-dopa or similar drugs, which heighten the dopamine load on an already fragile and stressed cholinergic system.

People with ordinary Parkinson's disease may be active and retain their intellectual powers for decades—Thomas Hobbes, the philosopher, for instance, developed "the shaking palsy" around the age of fifty, when he was completing his *Leviathan*, but remained intellectually intact and creative, though motorically disabled, into his nineties. But it has been increasingly recognized in the last few years that there is a more malignant form of parkinsonism, one accompanied sooner or later by dementia and by visual hallucinations even in the absence of L-dopa. Examination of the brain at autopsy in such patients may show abnormal aggregates of protein (so-called Lewy bodies) inside the nerve cells, mostly in the brain stem and basal ganglia but also in the visual association cortex. The Lewy bodies, it is conjectured, may predispose patients to visual hallucinations even before they are put on L-dopa.

Edna B. seems to have this disease, though the diagnosis of Lewy body disease cannot be made with certainty in life without doing a brain biopsy. Mrs. B. enjoyed excellent health until her mid-sixties, but in 2009 she developed some tremor in the hands, her first symptoms of parkinsonism. By the summer of 2010 her symptoms included some slowing of movement and speech as well as problems with memory and concentration—she would forget words and thoughts, lose the thread of what she was saying and thinking, and, most distressingly of all, she had hallucinations.

When I saw her in 2011, I asked her what her hallucinations

were like. "Horrible!" she said. "It's like watching a horror movie, and you're part of it." She saw little people ("Chuckys") running around her bed at night; they seemed to be talking to each other, she saw their gestures and their lips move, but she could not hear any speech. On one occasion she tried to speak to them. Although they looked frightening and (she thought) had evil intentions, they never molested or approached her, though once one of them sat on her bed. But far worse were certain scenes enacted before her. "I saw my son murdered right in front of my eyes," she told me. ("It was Darkside stuff," her husband interpolated.) Once, when her husband visited, she said, "What are you doing here? They just had your funeral at Sacred Heart Church." She often saw rats, and sometimes felt them in her bed. She also felt "fish" nibbling at her feet. Sometimes she had hallucinations of being part of an army marching into battle.

When I asked if she had any pleasant hallucinations, she said that she had sometimes seen people "in Hawaiian dress" in the corridor or outside her window, getting ready to play music for her, though she never actually heard any music. What she did hear, however, were various noises—especially the sound of running water. No voices. ("Good thing I didn't have those," she said, "or they'd think I was really crazy.") There have been some olfactory hallucinations, too: "people around me with different kinds of scents."

When her hallucinations started, Mrs. B. was understandably terrified, and took them for reality—"I did not even *know* the word 'hallucination,' " she said. Then she found herself more able to distinguish hallucinations from reality, but this did not prevent her from being frightened when they occurred. She always looked to her husband for reality testing; she would ask him whether *he* saw, heard, felt, or smelled some of the things she did. Sometimes she would have distortions of vision—her husband's face would be disfigured by a down-curving, sneering smile or occasionally his mouth would be upturned, "like a smiley face." A particularly strange and frightening hallucination occurred recently. There is a poster of a Native American chief hanging above her bed, and this came to life for Mrs. B. the other day; the chief stepped out of the frame and seemed to be standing in the bedroom. To reassure her, her husband waved his hands in front of the picture to dissipate the hallucination—and the chief seemed to disintegrate, but then she felt *she* was disintegrating, too. On another occasion, clothes in the bedroom "started walking around," and she had to have her husband shake a pair of jeans to show that it was just this, and nothing more.

Hallucinations may also occur in other types of dementia, including moderately advanced Alzheimer's disease, though less often than they do in Lewy body disease. In such cases, hallucinations may give rise to delusions, or they may stem from delusions. There may also be, in Alzheimer's or other types of dementia, delusions of duplication or misidentification. One patient of mine, sitting next to her husband on an airplane, suddenly saw him as "an imposter" who, she believed, had murdered her husband and was now trying to take his place. Another patient of mine, while she recognized the nursing home she was in by day, felt that she had been transferred to a cunning "duplicate" of the home each night. Sometimes psychoses can be centered on delusions of persecution, and occasionally these lead to violent behavior: one such patient assaulted a harmless neighbor, whom she felt was "spying" on her. Hallucinations in Alzheimer's disease, like those of Lewy body disease, are usually embedded in a complex matrix of sensory deceptions, confusion, disorientation, and delusions, and are rarely isolated, "pure" phenomena as in Charles Bonnet syndrome.

I worked for many years with the eighty deeply parkinsonian postencephalitic patients I described in *Awakenings*. Many of them had been "frozen" for decades, virtually immobilized by their disease. Once I got to know them well (after they had been enabled to move and talk by L-dopa), I found that perhaps a third of them had experienced visual hallucinations for years *before* L-dopa was introduced—hallucinations of a predominantly benign and sociable sort. I was not sure why they hallucinated in this way, but I thought it might be related to their isolation and social deprivation, their longing for the world—an attempt to provide a virtual reality, a hallucinatory substitute for the real world which had been taken from them.

Gertie C. had had a half-controlled hallucinosis for decades before she started on L-dopa—bucolic hallucinations of lying in a sunlit meadow or floating in a creek near her childhood home. This changed when she was given L-dopa, and her hallucinations assumed a social and sometimes sexual character. When she told me about this, she added, anxiously, "You surely wouldn't forbid a friendly hallucination to a frustrated old lady like me!" I replied that if her hallucinations had a pleasant and controllable character, they seemed rather a good idea under the circumstances. After this, the paranoid quality dropped away, and her hallucinatory encounters became purely amicable and amorous. She developed a humor and tact and control, never allowing herself a hallucination before eight

in the evening and keeping its duration to thirty or forty minutes at most. If her relatives stayed too late, she would explain
firmly but pleasantly that she was expecting "a gentleman
visitor from out of town" in a few minutes' time, and she felt
he might take it amiss if he was kept waiting outside. She now
receives love, attention, and invisible presents from a hallucinatory gentleman who visits faithfully each evening.

6

# Altered States

Humans share much with other animals—the basic needs of food and drink or sleep, for example—but there are additional mental and emotional needs and desires which are perhaps unique to us. To live on a day-to-day basis is insufficient for human beings; we need to transcend, transport, escape; we need meaning, understanding, and explanation; we need to see overall patterns in our lives. We need hope, the sense of a future. And we need freedom (or at least the illusion of freedom) to get beyond ourselves, whether with telescopes and microscopes and our ever-burgeoning technology or in states of mind which allow us to travel to other worlds, to transcend our immediate surroundings. We need detachment of this sort as much as we need engagement in our lives.

We may search, too, for a relaxing of inhibitions that makes it easier to bond with one another, or for transports that make our consciousness of time and mortality easier to bear. We seek a holiday from our inner and outer restrictions, a more intense sense of the here and now, the beauty and value of the world we live in.

William James was deeply interested, throughout his life, in the mystagogic powers of alcohol and other intoxicants, and he wrote about this in his 1902 book *The Varieties of Religious Experience.* He described, too, his own transcendent experiences with nitrous oxide:

> Our normal waking consciousness, rational consciousness, as we call it, is but one special type of consciousness, whilst all about it, parted from it by the filmiest of screens, there lie potential forms of consciousness entirely different. . . . Looking back on my own experiences, they all converge towards a kind of insight to which I cannot help ascribing some mystical significance. The keynote of it is invariably a reconciliation. It is as if the opposites of the world, whose contradictoriness and conflict make all our difficulties and troubles, were melted into unity. . . . To me [this sense] only comes in the artificial mystic state of mind.

Many of us find the reconciliation that James speaks of and even Wordsworthian "intimations of immortality" in nature, art, creative thinking, or religion; some people can reach transcendent states through meditation or similar trance-inducing techniques or through prayer and spiritual exercises. But drugs offer a shortcut; they promise transcendence on demand. These shortcuts are possible because certain chemicals can directly stimulate many complex brain functions.

Every culture has found chemical means of transcendence, and at some point the use of such intoxicants becomes institutionalized at a magical or sacramental level; the sacramental use of psychoactive plant substances has a long history and continues to the present day in various shamanic and religious rites around the world.

At a humbler level, drugs are used not so much to illuminate or expand or concentrate the mind, to "cleanse the doors of perception," but for the sense of pleasure and euphoria they can provide.

All of these cravings, high or low, are nicely met by the plant kingdom, which has various psychoactive agents that seem almost tailored to the neurotransmitter systems and receptor sites in our brains. (They are not, of course; they have evolved to deter predators or sometimes to attract other animals to eat a plant's fruit and disseminate its seeds. Nevertheless, one cannot repress a feeling of wonder that there should be so many plants capable of inducing hallucinations or altered brain states of many kinds.)[1]

Richard Evans Schultes, an ethnobotanist, devoted much of his life to the discovery and description of these plants and their uses, and Albert Hofmann was the Swiss chemist who first synthesized LSD-25 in a Sandoz lab in 1938. Together

---

[1.] Curiously, lower plants—cycads, conifers, ferns, mosses, and seaweeds— lack hallucinogenic substances.

Some nonflowering plants, however, contain stimulants, as the Mormons, among others, discovered. Mormons are forbidden to use tea or coffee. But on their long march along the Mormon Trail to Utah, the pioneers who were to found Salt Lake City, the new Zion, noticed a simple herb by the roadside, an infusion of which ("Mormon tea") refreshed and stimulated the weary pilgrims. The herb was ephedra, which contains ephedrine, chemically and pharmacologically akin to the amphetamines.

Schultes and Hofmann described nearly a hundred plants containing psychoactive substances in their *Plants of the Gods*, and new ones continue to be discovered (to say nothing of new compounds synthesized in the lab).[2]

M any people experiment with drugs, hallucinogenic and otherwise, in their teenage or college years. I did not try them myself until I was thirty and a neurology resident. This long virginity was not due to lack of interest.

I had read the great classics—De Quincey's *Confessions of an English Opium-Eater* and Baudelaire's *Artificial Paradises*, among others—at school. I had read about the French novelist Théophile Gautier, who in 1844 paid a visit to the recently founded Club des Hashischins, in a quiet corner of the Île Saint-Louis. Hashish, in the form of a greenish paste,

---

2. Quite by accident, Hofmann discovered the hallucinogenic powers of LSD when he synthesized a new batch of the chemical in 1943. He must have absorbed some through his fingertips, for later that day he began to feel odd and went home, thinking he had a cold. As he lay in bed, he experienced "an uninterrupted stream of fantastic images of extraordinary plasticity and vividness and accompanied by an intense kaleidoscopic play of colors." Jay Stevens, in his book *Storming Heaven: LSD and the American Dream*, recounted what came next:

> Suspecting that LSD-25 had caused these fireworks, Hofmann decided to test this hypothesis. . . . [A few days later] he dissolved what he thought was a prudently infinitesimal amount of the drug—250 millionths of a gram—in a glass of water and drank it down. [Forty minutes later] he recorded a growing dizziness, some visual disturbance, and a marked desire to laugh. Forty-two words later he stopped writing altogether and asked one of his lab assistants to call a doctor before accompanying him home. Then he climbed onto his bicycle—wartime shortages having made automobiles impractical—and pedaled off into a suddenly anarchic universe.

had recently been introduced from Algeria and was all the rage in Paris. At the salon, Gautier consumed a substantial piece of hash ("about as large as a thumb"). At first he felt nothing out of the ordinary, but soon, he wrote, "everything seemed larger, richer, more splendid," and then more specific changes occurred:

> An enigmatic personage suddenly appeared before me . . . his nose was bent like the beak of a bird, his green eyes, which he wiped frequently with a large handkerchief, were encircled with three brown rings, and caught in the knot of a high white starched collar was a visiting card which read: *Daucus-Carota, du Pot d'or.* . . . Little by little the salon was filled with extraordinary figures, such as are found only in the etchings of Callot or the aquatints of Goya; a *pêle-mêle* of rags and tatters, bestial and human shapes. . . . Singularly intrigued, I went straightaway to the mirror. . . . One would have taken me for a Javanese or Hindu idol: my forehead was high, my nose, lengthened into a trunk, curved onto my chest, my ears brushed my shoulders, and to make matters more discomforting still, I was the color of indigo, like Shiva, the blue deity.[3]

By the 1890s, Westerners were also beginning to sample mescal, or peyote, previously used only as a sacrament in certain Native American traditions.[4]

---

3. I am quoting from the translation provided by David Ebin in his excellent book *The Drug Experience: First-Person Accounts of Addicts, Writers, Scientists, and Others.*

4. Louis Lewin, a German pharmacologist, published the first scientific analysis of the peyote cactus in 1886, and it was named *Anhalonium lewinii* in his honor. Later, he sought to classify various psychoactive substances based on their pharmacological effects, and he divided them

As a freshman at Oxford, free to roam the shelves and stacks of the Radcliffe Science Library, I read the first published accounts of mescal, including ones by Havelock Ellis and Silas Weir Mitchell. They were primarily medical men, not just literary ones, and this seemed to lend an extra weight and credibility to their descriptions. I was captivated by Weir Mitchell's dry tone and his nonchalance about taking what was then an unknown drug with unknown effects.

At one point, Mitchell wrote in an 1896 article for the *British Medical Journal*, he took a fair portion of an extract made from mescal buttons and followed this up with four further doses. Although he noted that his face was flushed, his pupils were dilated, and he had "a tendency to talk, and now and then . . . misplaced a word," he nevertheless went out on house calls and saw several patients. Afterward, he sat down quietly in a dark room and closed his eyes, whereupon he experienced "an enchanted two hours," full of chromatic effects:

Delicate floating films of colour—usually delightful neutral purples and pinks. These came and went—now here, now there. Then an abrupt rush of countless points of white light swept across the field of view, as if the unseen millions of the Milky Way were to flow a sparkling river before the eye. In a minute this was over and the field was dark. Then I began to see zigzag lines of very bright colours, like those seen in some megrims [migraines]. . . . It was in rapid, what I might call

into five general groups: euphoriants or sedatives (like opium), inebriants (like alcohol), hypnotics (like chloral and kava), excitants (like amphetamine and coffee), and hallucinogens, which he called phantastica. Many drugs, he noted, had overlapping and paradoxical effects, so that stimulants or sedatives could sometimes be as hallucinogenic as peyote.

> minute, motion. . . . A white spear of grey stone grew up to huge height, and became a tall, richly finished Gothic tower of very elaborate and definite design. . . . As I gazed every projecting angle, cornice, and even the face of the stones at their joinings were by degrees covered or hung with clusters of what seemed to be huge precious stones, but uncut, some being more like masses of transparent fruit. These were green, purple, red, and orange. . . . All seemed to possess an interior light, and to give the faintest idea of the perfectly satisfying intensity and purity of these gorgeous colour-fruits is quite beyond my power. All the colours I have ever beheld are dull as compared to these.

He found he had no power to influence his visions voluntarily; they seemed to come at random or to follow some logic of their own.

Just as the introduction of hashish in the 1840s had led to a vogue for it, so these first descriptions of mescal's effects by Weir Mitchell and others in the 1890s and the ready availability of mescaline led to another vogue—for mescal promised an experience not only richer, longer-lasting, and more coherent than that induced by hashish, but one with the added promise of transporting one to mystical realms of unearthly beauty and significance.

Unlike Mitchell, who had focused on the colored, mostly geometric hallucinations that he compared in part to those of migraine, Aldous Huxley, writing of mescaline in the 1950s, focused on the transfiguration of the visual world, its investment with luminous, divine beauty and significance. He compared such drug experiences to those of great visionaries and artists, though also to the psychotic experiences of some schizophrenics. Both genius and madness, Huxley hinted, lay in these extreme states of mind—a thought not so different from those expressed by De Quincey, Coleridge, Baudelaire, and Poe in relation to their own ambiguous experiences with opium and hashish (and explored at length in Jacques Joseph Moreau's 1845 book *Hashish and Mental Illness*). I read Huxley's *Doors of Perception* and *Heaven and Hell* when they came out in the 1950s, and I was especially excited by his speaking of the "geography" of the imagination and its ultimate realm—the "Antipodes of the mind."[5]

Around the same time, I came across a pair of books by the physiologist and psychologist Heinrich Klüver. In the first one, *Mescal*, he reviewed the world literature on the effects of mescal and described his own experiences with it. Keeping his eyes closed, as Weir Mitchell had done, he saw complex geometrical patterns:

Transparent oriental rugs, but infinitely small . . . plastic filigreed spherical objets d'art [like] radiolaria . . . wallpaper

---

5. Benny Shanon uses this phrase as the title of his remarkable book *The Antipodes of the Mind*, which is based on personal experience as well as extensive cultural and anthropological experience with the South American hallucinogen ayahuasca. Ayahuasca is, in fact, a blend of two plants: *Psychotria viridis* and *Banisteriopsis caapi*, neither of which has any hallucinogenic power by itself. The leaves of *Psychotria* contain dimethyltryptamine (DMT), a very powerful hallucinogen—but DMT, if taken by mouth, is deactivated in the gut by monoamine oxidase (MAO). *Banisteriopsis*, however, contains compounds that inhibit the MAO and so allow the DMT to be absorbed. "When one thinks about it," Shanon writes, "the discovery of Ayahuasca is indeed amazing. The number of plants in the rain forest is enormous, the number of their possible pairings astronomical. The common sense method of trial and error would not seem to apply."

designs . . . cobweb-like figures or concentric circles and
squares . . . architectural forms, buttresses, rosettes, leafwork,
fretwork.

For Klüver these hallucinations represented an abnormal activation in the visual system, and he observed that
similar hallucinations could occur in a variety of other
conditions—migraine, sensory deprivation, hypoglycemia,
fever, delirium, or the hypnagogic and hypnopompic states
that come immediately before and after sleep. In *Mechanisms
of Hallucination*, published in 1942, Klüver spoke of the tendency to "geometrization" in the brain's visual system, and he
regarded all such geometrical hallucinations as permutations
of four fundamental "form constants" (he identified these as
lattices, spirals, cobwebs, and tunnels). He implied that such
constants must reflect something about the organization, the
functional architecture, of the visual cortex—but there was
little more to be said about this in the 1940s.

It might be said that both approaches—the "high," mystical approach of Huxley and the "low," neurophysiological
approach of Klüver—were too narrowly focused and failed
to do justice to the range and complexity of the phenomena
that mescaline could induce. This became clearer in the late
1950s, when LSD, as well as psilocybin mushrooms and morning glory seeds (both of which contain LSD-like compounds),
became widely available, ushering in a new hallucinogenic
drug age and a new word to go with it: "psychedelic."

Daniel Breslaw, a young man just out of college in the 1960s,
was one of the subjects in a study of LSD at Columbia University, and he gave a vivid description of the effects of psilocybin,
which he took under supervision, so that his reactions could

be observed.[6] His first visions, like Weir Mitchell's, were of stars and colors:

> I closed my eyes. "I see stars!" I then burst out, finding the firmament spread out on the inside of my eyelids. The room about me receded into a tunnel of oblivion as I vanished into another world, fruitless to describe. . . . The heavens above me, a night sky spangled with eyes of flame, dissolve into the most overpowering array of colors I have ever seen or imagined; many of the colors are entirely new—areas of the spectrum which I seem to have hitherto overlooked. The colors do not stand still, but move and flow in every direction; my field of vision is a mosaic of unbelievable complexity. To reproduce an instant of it would involve years of labor, that is, if one were able to reproduce colors of equivalent brilliance and intensity.

Then Breslaw opened his eyes. "With the eyes closed," he noted, "one is *not here*, but inhabits a distant world of abstractions. But with eyes open, one glances around the physical universe with curiosity." Curiosity—and amazement, for the visual world he saw was bizarrely changed and continually changing, as Gautier had found with hashish. Breslaw wrote:

> The room is fifty feet tall. Now it is two feet tall. A strange disparity here. Whatever comes into the focus of my eyes dissolves into whorls, patterns, arrangements. There is The Doctor. His face is crawling with lice. His glasses are the size

---

6. Breslaw's account is included in David Ebin's book *The Drug Experience.*

of pressure cookers, and his eyes are those of some mammoth fish. He is beyond doubt the funniest sight I have ever seen, and I insist upon this point by laughing. . . . A footstool in the corner shrinks to a mushroom in jerky spasms, braces—and springs to the ceiling. Amazing! . . . In the elevator, the face of the operator grows hair, becomes an affably growing gorilla.

Time was immensely distended. The elevator descended, "passing a floor every hundred years. Back in the room, I swim through the remaining centuries of the day. Every five eons or so a nurse arrives (in the aspect of a cougar, a differential equation, or a clock radio) and takes my blood pressure."

Animation and intentionality appeared everywhere, as did relationship and meaning:

Here is a fire extinguisher in a glass case, evidently an exhibit of some sort. A bit of staring reveals that the beast is alive: it coils its rubber hose around its prey and sucks flesh through the nozzle. The beast and I exchange glares, and then the nurse drags me away. I wave goodbye.

A smudge on the wall is an object of limitless fascination, multiplying in size, complexity, color. But more than that, one sees *every relationship it has to the rest of the universe*; it possesses, therefore, an endless variety of meanings, and one proceeds to entertain every possible thought there is to think about it.

And when the effects were most intense, there came a rich synesthesia—a mingling of all the senses, and of sensation and concepts. Breslaw noted, "Interchanges between the senses are frequent and astonishing: One knows the smell of a low B flat, the sound of green, the taste of the categorical imperative (which is something like veal)."

No two people ever have the same responses to such drugs; indeed, no two drug experiences are ever the same for the same person. Eric S. wrote to me to describe some of his experiences with LSD during the 1970s:

> I was in my late twenties when a friend and I took some LSD. I had tripped many times before but this acid was different. . . . We noticed that we were talking to each other mentally through thoughts only, no verbal talk, tele-communicating. I thought in my head, "I want a beer," and he heard me and got me a beer; he thought, "Turn the music up" and I turned the music up. . . . It went on like this for some time.
>
> Then I went to urinate, and in my urine stream was a video or movie of the past played back in reverse. Everything that had just happened in the room was coming out of me like watching a movie in my urine stream, playing in reverse. This totally blew my mind.
>
> Then my eyes became a microscope, and I looked at my wrist and was able to see each individual cell breathing or respirating, like little factories with little puffs of gas shooting out of each cell, some blowing perfect smoke rings. My eyes were able to see inside each skin cell, and I saw that I was choking myself from the inside by smoking five packs of cigarettes a day and the debris was clogging my cells. At that second I quit smoking.
>
> Then I left my body and hovered in the room above the whole scene, then found myself traveling through a tunnel of beautiful light into space and was filled with a feeling of total love and acceptance. The light was the most beautiful,

warm and inviting light I ever felt. I heard a voice ask me if I wanted to go back to Earth and finish my life or . . . to go in to the beautiful love and light in the sky. In the love and light was every person that ever lived. Then my whole life flashed in my mind from birth to the present, with every detail that ever happened, every feeling and thought, visual and emotional was there in an instant. The voice told me that humans are "Love and Light." . . .

That day will live with me forever; I feel I was shown a side of life that most people can't even imagine. I feel a special connection to every day, that even the simple and mundane have such power and meaning.

The effects of cannabis, mescaline, LSD, and other hallucinogenic drugs have an immense range and variety. Yet certain categories of perceptual distortion and hallucinatory experience may, to some extent, be regarded as typical of the brain's responses to such drugs.

The experience of color is often heightened, sometimes to an unearthly level, as Weir Mitchell, Huxley, and Breslaw all observed. There may be sudden changes in orientation and striking alterations of apparent size. There may be micropsia or Lilliputian vision (little beings—elves, dwarfs, fairies, imps— are curiously common in these hallucinations), or there may be gigantism (macropsia).

There may be exaggerations or diminutions of depth and perspective or exaggerations of stereo vision—or even stereo hallucinations, seeing three-dimensional depth and solidity in a flat picture. Huxley described this:

I was handed a large colored reproduction of the well-known self-portrait by Cézanne—the head and shoulders of a man in a large straw hat, red-cheeked, red-lipped, with rich black whiskers and a dark unfriendly eye. It is a magnificent painting; but it was not as a painting that I now saw it. For the head promptly took on a third dimension and came to life as a small goblin-like man looking out through a window in the page before me.

The perceptual transformations and hallucinations induced by mescaline, LSD, and other hallucinogens are predominantly, but not exclusively, visual. There may be enhancements or distortions or hallucinations of taste and smell, touch and hearing, or there may be fusions of the senses—a sort of temporary synesthesia—"the smell of a low B flat, the sound of green," as Breslaw put it. Such coalescences or associations (and their presumed neural basis) are creations of the moment. In this way they are quite different from true synesthesia, a congenital (and often familial) condition where there are fixed sensory equivalences that last a lifetime. With hallucinogens, time may appear to be distended or compressed. One may cease to perceive motion as continuous and see instead a series of static "snapshots," as with a film run too slowly. Such stroboscopic or cinematic vision is a not uncommon effect of mescaline. Sudden accelerations, slowings, or freezings of movement are also common with more elementary hallucinatory patterns.[7]

---

7. I have discussed neurological aspects of time and motion perception, as well as cinematic vision, at greater length in two articles, "Speed" and "In the River of Consciousness."

I had done a great deal of reading, but had no experiences of my own with such drugs until 1953, when my childhood friend Eric Korn came up to Oxford. We read excitedly about Albert Hofmann's discovery of LSD, and we ordered 50 micrograms of it from the manufacturer in Switzerland (it was still legal in the mid-1950s). Solemnly, even sacramentally, we divided it and took 25 micrograms each—not knowing what splendors or horrors awaited us—but, sadly, it had absolutely no effect on either of us. (We should have ordered 500 micrograms, not 50.)

By the time I qualified as a doctor, at the end of 1958, I knew I wanted to be a neurologist, to study how the brain embodies consciousness and self and to understand its amazing powers of perception, imagery, memory, and hallucination. A new orientation was entering neurology and psychiatry at that time; it was the opening of a neurochemical age, with a glimpse of the range of chemical agents, neurotransmitters, which allow nerve cells and different parts of the nervous system to communicate with one another. In the 1950s and 1960s, discoveries were coming from all directions, though it was far from clear how they fit together. It had been found, for instance, that the parkinsonian brain was low in dopamine, and that giving a dopamine precursor, L-dopa, could alleviate the symptoms of Parkinson's disease, while tranquilizers, introduced in the early 1950s, could depress dopamine and cause a sort of chemical parkinsonism. For about a century, the staple medication for parkinsonism had been anticholinergic drugs. How did the dopamine and acetylcholine systems interact? Why did opiates—or cannabis—have such strong effects? Did the brain have special opiate receptors and make opioids of its own? Was there a similar mechanism for cannabis receptors and cannabinoids? Why was LSD so enormously potent? Were all its effects explicable in terms of altering the serotonin in the brain? What transmitter systems governed wake-sleep cycles, and what might be the neurochemical background of dreams or hallucinations?

Starting a neurology residency in 1962, I found the atmosphere heady with such questions. Neurochemistry was plainly "in," and so—dangerously, seductively, especially in California, where I was studying—were the drugs themselves.

While Klüver had little idea of what the neural basis of his hallucinatory constants might be, rereading his book in the early 1960s was especially exciting to me in light of the groundbreaking experiments on visual perception that David Hubel and Torsten Wiesel were performing at the time, recording from neurons in the visual cortex in animals. They described neurons specialized for the detection of lines, orientations, edges, corners, etc., and these, it seemed to me, if stimulated by a drug or a migraine or a fever, might well produce just such geometrical hallucinations as Klüver had described.

But mescal hallucinations did not stop with geometrical designs. What was happening in the brain when one hallucinated more complex things: objects, places, figures, faces—let alone the heaven and hell that Huxley had described? Did *they* have their own basis in the brain?[8]

---

8. Very little was known in the early 1960s about how psychoactive drugs worked, and early research by Timothy Leary and others at Harvard, as well as the work of L. Jolyon West and Ronald K. Siegel at UCLA in the 1970s, focused mostly on the experiences of hallucinogens rather than their mechanisms. In 1975, Siegel and West published a wide-ranging collection of essays in their book *Hallucinations: Behavior, Experience, and Theory*. Here West set out (as he had in previous work) his release theory of hallucination.

It is now known that stimulants like cocaine and the amphetamines

Thoughts like this tipped the balance for me, along with the feeling that I would never really know what hallucinogenic drugs were like unless I tried them.

I started with cannabis. A friend in Topanga Canyon, where I lived at the time, offered me a joint; I took two puffs and was transfixed by what happened then. I gazed at my hand, and it seemed to fill my visual field, getting larger and larger while at the same time moving away from me. Finally, it seemed to me, I could see a hand stretched across the universe, light-years or parsecs in length. It still looked like a living, human hand, yet this cosmic hand somehow also seemed like the hand of God. My first pot experience was marked by a mix of the neurological and the divine.

On the West Coast in the early 1960s, LSD and morning glory seeds were readily available, so I sampled those, too. "But if you want a really far-out experience," my friends on Muscle Beach told me, "try Artane." I found this surprising, for I knew that Artane, a synthetic drug allied to belladonna, was used in modest doses (two or three tablets a day) for the treatment of Parkinson's disease, and that such drugs, in large quantities, could cause a delirium (such deliria have long been observed with accidental ingestion of plants like deadly nightshade, thorn apple, and black henbane). But would a delirium be fun? Or informative? Would one be in a position to observe the aberrant functioning of one's brain—to appreciate its wonder?

---

stimulate the "reward systems" of the brain, which are largely mediated by the neurotransmitter dopamine; this is also the case with opiates and alcohol. The classical hallucinogens—mescaline, psilocybin, LSD, and probably DMT—act by boosting serotonin in the brain.

"Go on," urged my friends. "Just take twenty of them—you'll still be in partial control."

So one Sunday morning, I counted out twenty pills, washed them down with a mouthful of water, and sat down to await the effect. Would the world be transformed, newborn, as Huxley had described it in *The Doors of Perception*, and as I myself had experienced with mescaline and LSD? Would there be waves of delicious, voluptuous feeling? Would there be anxiety, disorganization, paranoia? I was prepared for all of these, but none of them occurred. I had a dry mouth, large pupils, and found it difficult to read, but that was all. There were no psychic effects whatever—most disappointing. I did not know exactly what I had expected, but I had expected *something.*

I was in the kitchen, putting on a kettle for tea, when I heard a knocking at my front door. It was my friends Jim and Kathy; they would often drop round on a Sunday morning. "Come in, door's open," I called out, and as they settled themselves in the living room, I asked, "How do you like your eggs?" Jim liked them sunny side up, he said. Kathy preferred them over easy. We chatted away while I sizzled their ham and eggs—there were low swinging doors between the kitchen and the living room, so we could hear each other easily. Then, five minutes later, I shouted, "Everything's ready," put their ham and eggs on a tray, walked into the living room—and found it completely empty. No Jim, no Kathy, no sign that they had ever been there. I was so staggered I almost dropped the tray.

It had not occurred to me for an instant that Jim and Kathy's voices, their "presences," were unreal, hallucinatory. We had had a friendly, ordinary conversation, just as we usually had. Their voices were the same as always; there had been no hint, until I opened the swinging doors and found the living room empty, that the whole conversation, at least their side of it, had been completely invented by my brain.

I was not only shocked, but rather frightened, too. With LSD and other drugs, I knew what was happening. The world would look different, feel different; there would be every characteristic of a special, extreme mode of experience. But my "conversation" with Jim and Kathy had no special quality; it was entirely commonplace, with nothing to mark it as a hallucination. I thought about schizophrenics conversing with their "voices," but typically the voices of schizophrenia are mocking or accusing, not talking about ham and eggs and the weather.

"Careful, Oliver," I said to myself. "Take yourself in hand. Don't let this happen again." Sunk in thought, I slowly ate my ham and eggs (Jim and Kathy's, too) and then decided to go down to the beach, where I would see the real Jim and Kathy and all my friends, and enjoy a swim and an idle afternoon.

I was pondering all this when I became conscious of a whirring noise above me. It puzzled me for a moment, and then I realized it was a helicopter preparing to descend, and that it contained my parents, who, wanting to make a surprise visit, had flown in from London and, arriving in Los Angeles, had chartered a helicopter to bring them to Topanga Canyon. I rushed into the bathroom, had a quick shower, and put on a clean shirt and pants—the most I could do in the three or four minutes before they arrived. The throb of the engine was almost deafeningly loud, so I knew that the helicopter must have landed on the flat rock beside my house. I rushed out, excitedly, to greet my parents—but the rock was empty, there was no helicopter in sight, and the huge pulsing noise of its engine had abruptly cut off. The silence and emptiness, the disappointment, reduced me to tears. I had been so joyfully excited, and now there was nothing at all.

I went back into the house and had put on the kettle for another cup of tea when my attention was caught by a spider on the kitchen wall. As I drew nearer to look at it, the spider called out, "Hello!" It did not seem at all strange to me that a spider should say hello (any more than it seemed strange to Alice when the White Rabbit spoke). I said, "Hello, yourself," and with this we started a conversation, mostly on rather technical matters of analytic philosophy. Perhaps this direction was suggested by the spider's opening comment: did I think that Bertrand Russell had exploded Frege's paradox? Or perhaps it was its voice—pointed, incisive, and just like Russell's voice (which I had heard on the radio, but also—hilariously—as it had been parodied in *Beyond the Fringe*).[9]

D uring the week, I would avoid drugs, working as a resident at UCLA's neurology department. I was amazed and moved, as I had been as a medical student in London, by the range of patients' neurological experiences, and I found that I could not comprehend these sufficiently, or come to terms with them emotionally, unless I attempted to describe or transcribe them. It was then that I wrote my first published papers and my first book (it was never published, because I lost the manuscript).

But on the weekends, I often experimented with drugs. I recall vividly one episode in which a magical color appeared to

---

9. When, decades later, I told this story to my friend Tom Eisner, an entomologist, I mentioned the spider's philosophical tendencies and Russellian voice. He nodded sagely and said, "Yes, I know the species."

me. I had been taught, as a child, that there were seven colors in the spectrum, including indigo (Newton had chosen these, somewhat arbitrarily, by analogy with the seven notes of the musical scale). But some cultures recognize only five or six spectral colors, and few people agree as to what indigo is like.

I had long wanted to see "true" indigo, and thought that drugs might be the way to do this. So one sunny Saturday in 1964, I developed a pharmacologic launchpad consisting of a base of amphetamine (for general arousal), LSD (for hallucinogenic intensity), and a touch of cannabis (for a little added delirium). About twenty minutes after taking this, I faced a white wall and exclaimed, "I want to see indigo now—*now*!"

And then, as if thrown by a giant paintbrush, there appeared a huge, trembling, pear-shaped blob of the purest indigo. Luminous, numinous, it filled me with rapture: It was the color of heaven, the color, I thought, which Giotto had spent a lifetime trying to get but never achieved—never achieved, perhaps, because the color of heaven is not to be seen on earth. But it had existed once, I thought—it was the color of the Paleozoic sea, the color the ocean used to be. I leaned toward it in a sort of ecstasy. And then it suddenly disappeared, leaving me with an overwhelming sense of loss and sadness that it had been snatched away. But I consoled myself: Yes, indigo *exists*, and it can be conjured up in the brain.

For months afterward, I searched for indigo. I turned over little stones and rocks near my house, looking for it. I examined specimens of azurite in the natural history museum—but even they were infinitely far from the color I had seen. And then, in 1965, when I had moved to New York, I went to a concert in the Egyptology gallery of the Metropolitan Museum of Art. In the first half, a Monteverdi piece was performed, and I was utterly transported. I had taken no drugs, but I felt a glorious river of music, four hundred years long, flowing from Monteverdi's mind into my own. In this ecstatic mood, I wandered out during the intermission and looked at the ancient Egyptian objects on display—lapis lazuli amulets, jewelry, and so forth—and I was enchanted to see glints of indigo. I thought: Thank God, it really exists!

During the second half of the concert, I got a bit bored and restless, but I consoled myself, knowing that I could go out and take a "sip" of indigo afterward. It would be there, waiting for me. But when I went out to look at the gallery after the concert was finished, I could see only blue and purple and mauve and puce—no indigo. That was nearly fifty years ago, and I have never seen indigo again.

When a friend and colleague of my parents'—Augusta Bonnard, a psychoanalyst—came to Los Angeles for a year's sabbatical in 1964, it was natural that we should meet. I invited her to my little house in Topanga Canyon, and we had a genial dinner together. Over coffee and cigarettes (Augusta was a chain-smoker; I wondered if she smoked even during analytic sessions), her tone changed, and she said, in her gruff, smoke-thickened voice, "You need help, Oliver. You're in trouble."

"Nonsense," I replied. "I enjoy life. I have no complaints; all is well in work and love." Augusta let out a skeptical grunt, but she did not push the matter further.

I had started taking LSD at this point, and if that was not available, I would take morning glory seeds instead (this was before morning glory seeds were treated with pesticides, as they are now, to prevent drug abuse). Sunday mornings were usually my drug time, and it must have been two or three months after meeting Augusta that I took a hefty dose of Heavenly Blue morning glory seeds. The seeds were jet black and of agate-like hardness, so I pulverized them with a pestle and mortar and then mixed them with vanilla ice cream. About twenty minutes after eating this, I felt intense nausea, but when it subsided, I found myself in a realm of paradisiacal stillness and beauty, a realm outside time, which was rudely broken into by a taxi grinding and backfiring its way up the steep trail to my house. An elderly woman got out of the taxi, and, galvanized into action, I ran towards her, shouting, "I know who you are—you are a *replica* of Augusta Bonnard. You look like her, you have her posture and movements, but you are *not* her. I am not deceived for a moment." Augusta raised her hands to her temples and said, "Oy! This is worse than I realized." She got back into the taxi, and took off without another word.

We had plenty to talk about the next time we met. My failure to recognize her, my seeing her as a "replica," she thought, was a complex form of defense, a dissociation which could only be called psychotic. I disagreed and maintained that my seeing her as a duplicate or impostor was neurological in origin, a disconnection between perception and feelings. The ability to identify (which was intact) had not been accompanied by the appropriate feeling of warmth and familiarity, and it was this contradiction which had led to the logical though absurd conclusion that she was a "duplicate." (This syndrome, which can occur in schizophrenia, but also with dementia or delirium, is known as Capgras syndrome.) Augusta said that whichever view was correct, taking mind-altering drugs every weekend, alone, and in high doses, surely testified to some intense inner needs or conflicts, and that I should explore these with a therapist. (In retrospect, I am sure she was right, and I began seeing an analyst a year later.)

The summer of 1965 was a sort of in-between time: I had completed my residency at UCLA and had left California, but I had three months ahead of me before taking up a research fellowship in New York. This should have been a time of delicious freedom, a wonderful and needed holiday after the sixty- and sometimes eighty-hour workweeks I had had at UCLA. But I did not feel free; I get unmoored, have a sense of emptiness and structurelessness, when I am not working—it was weekends which were the danger times, the drug times, when I lived in California—and now an entire summer in my hometown, London, stretched before me like a three-month-long weekend.

It was during this idle, mischievous time that I descended deeper into drug taking, no longer confining it to weekends. I tried intravenous injection, which I had never done before. My parents, both physicians, were away, and, having the house to myself, I decided to explore the drug cabinet in their surgery on the ground floor for something special to celebrate my thirty-second birthday. I had never taken morphine or any opiates before. I used a large syringe—why bother with piddling doses? And after settling myself comfortably in bed, I drew up the contents of several vials, plunged the needle into a vein, and injected the morphine very slowly.

Within a minute or so, my attention was drawn to a sort of commotion on the sleeve of my dressing gown, which hung on the door. I gazed intently at this, and as I did so, it resolved itself into a miniature but microscopically detailed battle scene. I could see silken tents of different colors, the largest of which was flying a royal pennant. There were gaily caparisoned horses, soldiers on horseback, their armor glinting in the sun, and men with longbows. I saw pipers with long silver pipes, raising these to their mouths, and then, very faintly, I heard their piping, too. I saw hundreds, thousands of men—two armies, two nations—preparing to do battle. I lost all sense of this being a spot on the sleeve of my dressing gown, of the fact that I was lying in bed, that I was in London, that it was 1965. Before shooting up the morphine, I had been reading Froissart's *Chronicles* and *Henry V*, and now these became conflated in my hallucination. I realized that what I was gazing at from my aerial viewpoint was Agincourt, late in 1415, that I was looking down on the serried armies of England and France drawn up to do battle. And in the great pennanted tent, I knew, was Henry V himself. I had no sense that I was imagining or hallucinating any of this; what I saw was actual, real.

After a while the scene started to fade, and I became dimly conscious, once more, that I was in London, stoned, hallucinating Agincourt on the sleeve of my dressing gown. It had been an enchanting and transporting experience, literally so, but now it was over. The drug effect was fading fast; Agincourt was hardly visible now. I glanced at my watch. I had injected the morphine at nine-thirty, and now it was ten. But I had a sense of something odd—it had been dusk when I took the morphine; it should be darker still. But it was not. It was getting lighter, not darker, outside. It *was* ten o'clock, I realized, but ten in the morning. I had been gazing, motionless, at my Agincourt for more than twelve hours. This shocked and sobered me, and made me realize that one could spend entire days, nights, weeks, even years of one's life in an opium stupor. I would make sure that my first opium experience was also my last.

At the end of that summer of 1965, I moved to New York to begin a postgraduate fellowship in neuropathology and neurochemistry. December 1965 was a bad time: I was finding New York difficult to adjust to after my years in California, a love affair had gone sour, my research was going badly, and I was discovering for myself that I was not cut out to be a bench scientist. Depressed and insomniac, I was taking ever-increasing amounts of chloral hydrate to get to sleep, and was up to fifteen times the usual dose every night. And though I had managed to stockpile a huge amount of the drug—I raided the chemical supplies in the lab at work—this finally ran out on a bleak Tuesday a little before Christmas, and for the first time in several months I went to bed without my usual knockout dose. My sleep was poor, broken by nightmares and bizarre dreams, and upon waking, I found myself excruciatingly sensitive to sounds. There were always trucks rumbling along the cobblestoned streets of the West Village; now it sounded as if they were crushing the cobblestones to powder as they passed.

Feeling a bit shaky, I did not ride my motorcycle to work as usual, but took a train and bus. Wednesday was brain-cutting day in the neuropathology department, and it was my turn to slice a brain into neat horizontal sections, to identify the main structures as I did so, and to observe whether there were any departures from normal. I was usually pretty good at this, but that day I found my hand trembling visibly, embarrassingly, and the anatomical names were slow in coming to mind.

When the session ended, I went across the road, as I often did, for a cup of coffee and a sandwich. As I was stirring the coffee, it suddenly turned green, then purple. I looked up, startled, and saw that a customer paying his bill at the cash register had a huge proboscidean head, like an elephant seal. Panic seized me; I slammed a five-dollar note on the table and ran across the road to a bus on the other side. But all the passengers on the bus seemed to have smooth white heads like giant eggs, with huge glittering eyes like the faceted compound eyes of insects—their eyes seemed to move in sudden jerks, which increased the feeling of their fearfulness and alienness. I realized that I was hallucinating or experiencing some bizarre perceptual disorder, that I could not stop what was happening in my brain, and that I had to maintain at least an external control and not panic or scream or become catatonic, faced by the bug-eyed monsters around me. The best way of doing this, I found, was to write, to describe the hallucination in clear, almost clinical detail, and, in so doing, become an observer, even an explorer, not a helpless victim of the craziness inside me. I am never without pen and notebook, and now I wrote for dear life, as wave after wave of hallucination rolled over me.

Description, writing, had always been my best way of dealing with complex or frightening situations—though it had never been tested in so terrifying a situation. But it worked; by describing what was going on in my lab notebook, I managed to maintain a semblance of control, though the hallucinations continued, mutating all the while.

I managed somehow to get off at the right bus stop and onto the train, even though everything was now in motion, whirling vertiginously, tilting and even turning upside down. And I managed to get off at the right station, in my neighborhood in Greenwich Village. As I emerged from the subway, the buildings around me were tossing and flapping from side to side, like flags blowing in a high wind. I was enormously relieved to make it back to my apartment without being attacked, or arrested, or killed by the rushing traffic on the way. As soon as I got inside, I felt I had to contact somebody—someone who knew me well, who was both a doctor and a friend. Carol Burnett was the person: we had interned together in San Francisco five years earlier and had resumed a close friendship now that we were both in New York City. Carol would understand; she would know what to do. I dialed her number with a grossly tremulous hand. "Carol," I said, as soon as she picked up, "I want to say good-bye. I've gone mad, psychotic, insane. It started this morning, and it's getting worse all the while."

"Oliver!" Carol said. "What have you just taken?"

"Nothing," I replied. "That's why I'm so frightened." Carol thought for a moment, then asked, "What have you just *stopped* taking?"

"That's it!" I said. "I was taking a huge amount of chloral hydrate and ran out of it last night."

"Oliver, you chump! You always overdo things," Carol said. "You've got a classic case of the DT's, delirium tremens."

This was an immense relief—much better DT's than a schizophrenic psychosis. But I was quite aware of the dangers of the DT's: confusion, disorientation, hallucination, delusion, dehydration, fever, rapid heartbeat, exhaustion, seizures, death. I would have advised anyone else in my state to get to an emergency room immediately, but for myself, I wanted to tough it out, and experience it to the full. Carol agreed to sit with me for the first day; then, if she thought I was safe by myself, she would look by or phone me at intervals, calling in outside help if she judged it necessary. Given this safety net, I lost much of my anxiety, and could even enjoy the phantasms of delirium tremens in a way (though the myriads of small animals and insects were anything but pleasant). The hallucinations continued for almost ninety-six hours, and when they finally stopped, I fell into an exhausted stupor.[10]

A s a boy, I had known extreme delight in the study of chemistry and the setting up of my own chemistry lab. This delight seemed to desert me at the age of fifteen or so; in my years at school, university, medical school, and then internship and residency, I kept my head above water, but the subjects I studied never excited me in the same intense way as chemistry had when I was a boy. It was not until I arrived in New York and began seeing patients in a migraine clinic in the summer of 1966 that I began to feel a little stirring of the intellectual excitement and emotional engagement I had known in my earlier years. It was in the hope of stirring up these intellectual and emotional excitements even further that I turned to amphetamines.

I would take the stuff on Friday evenings after getting back

---

10. Many years later, I experienced the much gentler effects of sakau, the intoxicating sap of a pepper (*Piper methysticum*, also called kava in Polynesia) cultivated in the South Pacific. Drinking sakau has been a central part of Micronesian life, as chewing coca leaves has been in the Andes, for thousands of years; and its use is formalized in elaborate sakau rituals. I described the effects of sakau at length in *The Island of the Colorblind*; it may evoke a delicious sense of floating and ease, as well as a variety of visual illusions or hallucinations.

from work and would then spend the whole weekend so high that images and thoughts would become rather like controllable hallucinations, imbued with ecstatic emotion. I often devoted these "drug holidays" to romantic daydreaming, but one Friday, in February 1967, while I was exploring the rare book section of the medical library, I found a hefty volume on migraine entitled *On Megrim, Sick-Headache, and Some Allied Disorders: A Contribution to the Pathology of Nerve-Storms*, written in 1873 by one Edward Liveing, MD. I had been working for several months in the migraine clinic, and I was fascinated by the range of symptoms and phenomena that could occur in migraine attacks. These attacks often included an aura, a prodrome in which aberrations of perception and even hallucinations occurred. They were entirely benign and would last only a few minutes, but those few minutes provided a window onto the functioning of the brain and how it could break down and then reintegrate. In this way, I felt, every attack of migraine opened out into an encyclopedia of neurology.

I had read dozens of articles about migraine and its possible basis, but none of them seemed to present the full richness of its phenomenology or the range and depth of suffering which patients might experience. It was in the hope of finding a fuller, deeper, and more human approach to migraine that I took out Liveing's book from the library that weekend. So, after downing my bitter draft of amphetamine—heavily sugared to make it more palatable—I started reading. As the amphetamine effect took hold of me, stimulating my emotions and imagination, Liveing's book seemed to increase in intensity and depth and beauty. I wanted nothing but to enter Liveing's mind and imbibe the atmosphere of the time in which he had worked.

In a sort of catatonic concentration so intense that in ten

hours I scarcely moved a muscle or wet my lips, I read steadily through the five hundred pages of *Megrim*. As I did so, it seemed to me almost as if I were becoming Liveing himself, actually seeing the patients he described. At times I was unsure whether I was reading the book or writing it. I felt myself in the Dickensian London of the 1860s and '70s. I loved Liveing's humanity and social sensitivity, his strong assertion that migraine was not some indulgence of the idle rich but could affect those who were poorly nourished and worked long hours in ill-ventilated factories. In this way, his book reminded me of Mayhew's great study of London's working classes, but equally, one could tell how well Liveing had been trained in biology and the physical sciences, and what a master of clinical observation he was. I found myself thinking, *This represents the best of mid-Victorian science and medicine; it is a veritable masterpiece!* The book gave me what I had been hungering for during the months that I had been seeing patients with migraine, frustrated by the thin, impoverished articles which seemed to constitute the modern "literature" on the subject. At the height of this ecstasy, I saw migraine shining like an archipelago of stars in the neurological heavens.

But a century had passed since Liveing worked and wrote in London. Rousing myself from my reverie of being Liveing or one of his contemporaries, I came to and said to myself, Now it is the 1960s, not the 1860s. Who could be the Liveing of our time? A disingenuous clutter of names spoke themselves in my mind. I thought of Dr. A. and Dr. B. and Dr. C. and Dr. D., all of them good men but none of them with that mix of science and humanism that was so powerful in Liveing. And then a very loud internal voice said, "You silly bugger! *You're* the man!"

On every previous occasion when I had come down after two days of amphetamine-induced mania, I had experienced a severe reaction in the other direction, feeling an almost narcoleptic drowsiness and depression. I would also have an acute sense of folly, thinking that I had endangered my life for nothing—amphetamines in the large doses I took would give me a sustained pulse rate close to 200 and a blood pressure of I know not what; several people I knew had died from overdoses of amphetamines. I would feel that I had made a crazy ascent into the stratosphere but had come back empty-handed and had nothing to show for it; that the experience had been as empty and vacuous as it was intense. This time, though, when I came down, I retained a sense of illumination and insight; I had had a sort of revelation about migraine. I had a sense of resolution, too, that I was indeed equipped to write a Liveing-like book, that perhaps *I* could be the Liveing of our time.

The next day, before I returned Liveing's book to the library, I photocopied the whole thing. Then, bit by bit, I started to write my own book. The joy I got from doing this was *real*—infinitely more substantial than the vapid mania of amphetamines—and I never took amphetamines again.

# 7

# Patterns:
# Visual Migraines

I have had migraines for most of my life; the first attack I remember occurred when I was three or four years old. I was playing in the garden when a shimmering light appeared to my left, dazzlingly bright. It expanded, becoming an enormous arc stretching from the ground to the sky, with sharp, glittering, zigzagging borders and brilliant blue and orange colors. Then behind the brightness came a growing blindness, an emptiness in the field of vision, and soon I could see almost nothing on my left side. I was terrified—what was happening? My sight returned to normal in a few minutes, but these were the longest minutes I had ever experienced.

I told my mother what had happened, and she explained to me that what I had had was a migraine aura—a feeling or sensation that precedes a migraine; she was a doctor, and she too was a "migraineur." It was a visual migraine aura, and the characteristic zigzag shape, she would later tell me, resembled that of medieval forts, so it was often called a fortification pattern. Many people, she said, would get a terrible headache after seeing the aura.

I was lucky to be one of those people who got only the aura without the headache, and lucky, too, to have a mother who could reassure me that everything would be back to normal within a few minutes, and with whom, as I got older, I could share my migraine experiences. She explained that auras like mine were due to a sort of electrical disturbance like a wave passing across the visual parts of the brain. A similar "wave" could pass over other parts of the brain, too, she said, so one might get a strange feeling on one side of the body or experience an odd smell or find oneself temporarily unable to speak. A migraine might affect one's perception of color or depth or movement, might make the whole visual world unintelligible for a few minutes. Then, if one were unlucky, the rest of the migraine would follow: violent headaches, vomiting, painful sensitivity to light and noise, abdominal disturbances, and a host of other symptoms.[1] Migraine was common, my mother said, affecting at least 10 percent of the population. Its classic visual presentation is a scintillating, zigzag-edged, kidney-shaped form like the one I saw, expanding and moving slowly across one half of the visual field over the course of fifteen or twenty minutes. Inside the shimmering borders of this

---

1. A migraine headache often occurs on only one side (hence the term, which derives from the Greek for "hemi" and "cranium"). But it can also be on both sides, and can range from a dull or throbbing ache to excruciating pain, as J. C. Peters described in his 1853 *A Treatise on Headache:*

> The character of the pains varied very much; most frequently they were of a hammering, throbbing or pushing nature. . . . [in other cases] pressing and dull . . . boring with sense of bursting . . . pricking . . . rending . . . stretching . . . piercing . . . and radiating. . . . In a few cases it felt as if a wedge was pressed into the head, or like an ulcer, or as if the brain was torn, or pressed outwards.

shape is often a blind area, a scotoma—thus the whole shape is called a scintillating scotoma.

For most people with classical migraine, the scintillating scotoma is the chief visual effect, and things go no further. But sometimes, within the scotoma, there are other patterns. In my own migraine auras, I would sometimes see—vividly with closed eyes, more faintly and transparently if I kept my eyes open—tiny branching lines like twigs or geometrical structures: lattices, checkerboards, cobwebs, and honeycombs. Unlike the scintillating scotoma itself, which had a fixed appearance and a slow, steady rate of progression, these patterns were in continual motion, forming and re-forming, sometimes assembling themselves into more complicated forms like Turkish carpets or complex mosaics or three-dimensional shapes like tiny pinecones or sea urchins. Usually these patterns stayed inside the scotoma, to one side or the other of my visual field, but sometimes they seemed to break loose and scatter themselves all over.

One has to call these hallucinations, even though they are only patterns and not images, for there is nothing in the external world that corresponds to the zigzags and checkerboards—they are generated by the brain. And there may also be startling perceptual changes with migraine. I might sometimes lose the sense of color or of depth (for other people, color or depth may intensify). Losing the sense of movement was especially startling, for instead of continuous movement, I would see only a stuttering series of "stills." Objects might change size or shape or distance, or get misplaced in the visual field so that, for a minute or two, the whole visual world would be unintelligible.

There are many variations on the visual experiences of migraine. Jesse R. wrote to me that during a migraine, "I think my mind loses its ability to read shapes and misinterprets them. . . . I think I see a person instead of the coat rack . . . or I often think I see movement across a table or floor. What is strange is that the mind always errs toward giving life to inanimate objects."

Toni P. wrote that before her migraines, she might see alternating black and white zigzag lines in her peripheral vision: "shiny geometric shapes, flashes of light. Sometimes it is as if whatever I am viewing is through a sheer curtain that is blowing in the wind." But sometimes, for her, a scotoma is simply a blank spot, producing an uncanny sense of nothingness:

> I was studying for a major lab exam when all of a sudden I knew something was missing—the book was in front of me; I could see the edges, but there were no words, no graphs, no diagrams. It wasn't as if there was a blank page, it just didn't exist. I only knew it SHOULD be there by reason. That was the strangeness of it. . . . It lasted for about twenty minutes.

Another woman, Deborah D., had an attack of migraine in which, she wrote:

> When I looked at the computer screen, I could not read anything; the screen was a crazy blur . . . of multiple images. . . . I could not see the numbers on the phone's keypad, it was as if I was seeing through a fly's lens, multiple images, not double, not triple, but many, many images of wherever I looked.

It is not only the visual world that may be affected in a migraine aura. There may be hallucinations of body image—the feeling that one is taller or shorter, that one limb has shrunk or grown gigantic, that one's body is tilted at an angle, and so forth.

It is known that Lewis Carroll had classical migraines, and it has been suggested (by Caro W. Lippman and others) that his migraine experiences may have inspired Alice in Wonderland's strange alterations of size and shape. Siri Hustvedt, in a *New York Times* blog, described her own transcendent Alice-in-Wonderland syndrome:

> As a child I had what I called "lifting feelings." Every once in a while I had a powerful internal sensation of being pulled upward, as if my head were rising, even though I knew my feet hadn't left the ground. This lift was accompanied by what can only be called awe—a feeling of transcendence. I variously interpreted these elevations as divine (God was calling) or as an amazed connection to things in the world. Everything appeared strange and wondrous.

There may be auditory misperceptions and hallucinations in migraine: sounds are amplified, reverberant, distorted; occasionally voices or music are heard. Time itself may seem distorted.

Hallucinations of smell are not uncommon—the smells are often intense, unpleasant, strangely familiar, yet unspecifiable. I myself twice hallucinated a smell before a migraine, but a pleasant one—the smell of buttered toast. The first time it happened, I was at the hospital and went in search of the toast—it did not occur to me that I was having a hallucination until the visual fortifications started up, a few minutes later. On both occasions there was a memory or pseudomemory of being a little boy in a high chair about to have buttered toast at teatime. One migraineur wrote to me, "I have always smelled beef roasting about thirty minutes before the onset of a migraine."[2] A patient described by G. N. Fuller and R. J. Guiloff had "vivid olfactory hallucinations, lasting five minutes, of either her grandfather's cigars or peanut butter."

When I worked in a migraine clinic as a young neurologist, I made a point of asking every patient about such experiences. They were usually relieved that I asked, for people are afraid to mention hallucinations, fearing that they will be seen as psychotic. Many of my patients habitually saw patterns in their migraine auras, and a few had a host of other strange visual phenomena, including distortion of faces or objects melting or flickering into one another; multiplication of objects or figures; or persistence or recurrence of visual images.

Most migraine auras remain at the level of elementary hallucinations: phosphenes, fortifications, and geometrical figures of other sorts—but more complex hallucinations, though rare in migraine, do occur. My colleague Mark Green, a neurologist, described to me how one of his patients had the same vision in every migraine attack: a hallucination of a worker emerging from a manhole in the street, wearing a white hard hat with an American flag painted on it.

S. A. Kinnier Wilson, in his encyclopedic *Neurology*, de-

---

2. This woman, Ingrid K., also reported that she sometimes has "another strange experience just before the migraine . . . I think I recognize everyone I see. I don't know who they are . . . but everyone looks familiar." Other correspondents have described a similar "hyperfamiliarity" at the start of migraine—and this feeling is occasionally part of an epileptic aura, as Orrin Devinsky and his colleagues have described.

scribed how a friend of his would always have a stereotyped hallucination as part of a migraine aura:

> [He] used at first to see a large room with three tall arched windows and a figure clad in white (its back toward him) seated or standing at a long bare table; for years this was the unvarying aura, but it was gradually replaced by a cruder form (circles and spirals), which, later still, developed once in a while without subsequent headache.

Klaus Podoll and Derek Robinson, in their beautifully illustrated monograph *Migraine Art*, have collected many reports of complex hallucinations in migraine from the world literature. People may see human figures, animals, faces, objects, or landscapes—often multiplied. One man reported seeing "a fly's eye made of millions of light-blue Mickey Mouses" during a migraine attack, but this hallucination was confined to the temporarily blind half of his visual field. Another saw a "crowd of [more than] one hundred people, some dressed in white."

There may also be lexical hallucinations. Podoll and Robinson cite a case from the nineteenth-century literature:

> A patient of Hoeflmayr's saw words written in the air; a patient of Schob's had hallucinations of letters, words, and numbers; and a patient reported by Fuller et al. "saw writing on the wall and when asked what it was said he was too far from it. He then walked up to the wall and was able to read it out clearly."

Lilliputian hallucinations can occur in migraine (as well as in other conditions), as Siri Hustvedt described in a *New York Times* blog:

I was lying in bed reading a book by Italo Svevo, and for some reason, looked down, and there they were: a small pink man and his pink ox, perhaps six or seven inches high. They were perfectly made creatures and, except for their color, they looked very *real*. They didn't speak to me, but they walked around, and I watched them with fascination and a kind of amiable tenderness. They stayed for some minutes and then disappeared. I have often wished they would return, but they never have.

All of these effects seem to show, by default, what a colossal and complicated achievement normal vision is, as the brain constructs a visual world in which color and movement and size and form and stability are all seamlessly meshed and integrated. I came to regard my own migraine experiences as a sort of spontaneous (and fortunately reversible) experiment of nature, a window into the nervous system—and I think this was one reason I decided to become a neurologist.

What is stirring up the visual system during a migraine attack, to provoke such hallucinations? William Gowers, writing more than a century ago, when little was known of the cellular details of the visual cortex (or the brain's electrical activity), addressed this question in *The Border-land of Epilepsy:*

The process which gives rise to the sensory symptoms . . . of migraine is very mysterious. . . . There is a peculiar form of activity which seems to spread, like the ripples in a pond into which a stone is thrown. But the activity is slow, deliberate, occupying twenty minutes or so in passing through the centre affected. In the region through which the active ripple waves have passed, a state is left like molecular disturbance of the structures.

Gowers's intuition proved quite accurate and was given physiological backing decades later, when it was discovered that a wave of electrical excitation could track across the cerebral cortex at much the same rate the fortifications did. In 1971, Whitman Richards suggested that the zigzag shape of migraine fortification patterns, with its characteristic angles, might reflect something equally constant in the architecture of the visual cortex itself—perhaps clusters of the orientation-sensitive neurons which Hubel and Wiesel had demonstrated in the early 1960s. As the wave of electrical excitation slowly marches across the cortex, Richards suggested, it might directly stimulate these clusters, causing the patient to "see" shimmering bars of light at different angles. But it was only with the use of magnetoencephalography, twenty years later, that it was possible to demonstrate that the passage of fortifications in a migraine aura was indeed accompanied by just such a wave of electrical excitation.

A hundred and fifty years ago, the astronomer Hubert Airy (who was a migraineur himself) felt that the aura of migraine provided "a sort of photograph" of the brain in action. He, like Gowers, may have been more literally accurate than he knew.

Heinrich Klüver, writing about mescal, remarked that the simple geometric hallucinations one might get with hallucinogenic drugs were identical to those found in migraine and many other conditions. Such geometrical forms, he felt, were not dependent on memory or personal experience or desire or imagination; they were built into the very architecture of the brain's visual systems.

But while the zigzag fortification patterns are highly stereotyped and can perhaps be understood in terms of the orientation receptors in the primary visual cortex, a different sort of explanation must be sought for the rapidly changing, permuting play of geometric forms. Here we need dynamic explanations, a consideration of the ways in which the activity of millions of nerve cells can produce complex and ever-changing patterns. We can actually see, through such hallucinations, something of the dynamics of a large population of living nerve cells and, in particular, the role of self-organization in allowing complex patterns of activity to emerge. Such activity operates at a basic cellular level, far beneath the level of personal experience. The hallucinatory forms are, in this way, physiological universals of human experience.

Perhaps such experiences are at the root of our human obsession with pattern and the fact that geometrical patterns find their way into our decorative arts. As a child, I was fascinated by the patterns in our house—the square colored floor tiles on the front porch, the small hexagonal ones in the kitchen, the herringbone pattern on the curtains in my room, the check pattern on my father's suit. When I was taken to the synagogue for services, I was more interested in the mosaics of tiny tiles on the floor than in the religious liturgy. And I loved the pair of antique Chinese cabinets in our drawing room, for embossed on their lacquered surfaces were designs of wonderful intricacy on different scales, patterns nested within patterns, all surrounded by clusters of tendrils and leaves. These geometric and scrolling motifs seemed somehow familiar to me, though it did not dawn on me until years later that this was because I had seen them in my own head, that these patterns resonated with my own inner experience of the intricate tilings and swirls of migraine.

Migraine-like patterns, indeed, can be found in Islamic art, in classical and medieval motifs, in Zapotec architecture, in the bark paintings of Aboriginal artists in Australia, in Acoma pottery, in Swazi basketry—in virtually every culture, going back tens of thousands of years. There seems to have been, throughout human history, a need to externalize and make art from these internal experiences, from the cross-hatchings of prehistoric cave paintings to the swirling psychedelic art of the 1960s. Do the arabesques and hexagons in our own minds, built into our brain organization, provide us with our first intimations of formal beauty?

There is an increasing feeling among neuroscientists that self-organizing activity in vast populations of visual neurons is a prerequisite of visual perception—that this is how seeing begins. Spontaneous self-organization is not restricted to living systems; one may see it in the formation of snow crystals, in the roilings and eddies of turbulent water, in certain oscillating chemical reactions. Here, too, self-organization can produce geometries and patterns in space and time very similar to what one may see in a migraine aura. In this sense, the geometrical hallucinations of migraine allow us to experience in ourselves not only a universal of neural functioning but a universal of nature itself.

8

# The "Sacred" Disease

E pilepsy affects a substantial minority of the population, occurs in all cultures, and has been recognized since the dawn of recorded history. It was known to Hippocrates as the sacred disease, a disorder of divine inspiration.[1] And yet in its major, convulsive form (the only form recognized until the nineteenth century), it has attracted fear, hostility, and cruel discrimination. It still carries a good deal of stigma today.

Epileptic attacks—often called seizures or fits—can take a dozen or more forms. These have in common a sudden onset (sometimes without any warning, but sometimes with a characteristic prodrome or aura) and a basis in a sudden, abnormal

---

1. When Hippocrates wrote "On the Sacred Disease," he was bowing to the then-popular notion of epilepsy's divine origin, but he dismisses this in his opening sentence: "The disease called sacred . . . appears to me no more sacred than other diseases, but has a natural cause from which it originates, like other affections."

electrical discharge in the brain. In generalized seizures, this discharge arises from both halves of the brain simultaneously. In a grand mal seizure, there is violent, convulsive movement of the muscles, biting of the tongue, and sometimes foaming at the mouth; there may also be a harsh, inhuman-sounding "epileptic cry." Within seconds, the person having a grand mal seizure will lose consciousness and fall to the ground (epilepsy was also called "the falling sickness"). Such attacks can be terrifying to see.

In a petit mal seizure, there is only a transient loss of consciousness—the person seems to be "absent" for a few seconds, but may then continue a conversation or a chess game without realizing, or anyone else realizing, that anything unusual has happened.

In contrast to such generalized seizures, which arise from an inborn, genetic sensitivity of the brain, partial seizures arise from a particular area of damage or sensitivity in one part of the brain, an epileptic focus, which may be congenital or the result of an injury. The symptoms of partial seizures depend on the location of the focus: they may be motor (twitching of certain muscles), autonomic (nausea, a rising feeling in the stomach, etc.), sensory (abnormalities or hallucinations of sight, sound, smell, or other sensations), or psychic (sudden feelings of joy or fear without apparent cause, déjà vu or jamais vu, or sudden, often unusual, trains of thought). Partial seizure activity may be confined to the epileptic focus, or it may spread to other areas of the brain, and occasionally it leads to a generalized convulsion.

Partial or focal seizures were only recognized in the second half of the nineteenth century—a time when focal deficits of all kinds (for instance, aphasia, the loss of linguistic ability, or agnosia, the loss of ability to identify objects) were being described and attributed to damage in specific areas of the brain. This correlation of cerebral pathology with specific deficits, or "negative" symptoms, led to the understanding that there are many different centers in the brain crucial to certain functions.

But Hughlings Jackson (sometimes called the father of English neurology) paid equal attention to the "positive" symptoms of neurological disease—symptoms of overactivity, such as seizures, hallucinations, and deliria. He was a minute and patient observer, and he was the first to recognize "reminiscence" and "dreamy states" in complex seizures. We still speak of focal motor seizures which start in the hands and "march" up the arm as Jacksonian epilepsy.

Jackson was also an extraordinary theorist, who proposed that higher and higher levels had evolved in the human nervous system—and that these were hierarchically organized, with higher centers constraining lower ones. Thus, he thought, damage in the higher centers might cause "release" activity in the lower ones. For Jackson, epilepsy was a window into the organization and workings of the nervous system (as migraine was for me). "He who is faithfully analyzing many different cases of epilepsy," Jackson wrote, "is doing far more than studying epilepsy."

Jackson's younger partner in the enterprise of describing and classifying seizures was William Gowers, and where Jackson's writing was complex, convoluted, and full of reservations,

Gowers's was simple, transparent, and lucid. (Jackson never wrote a book, but Gowers wrote many, including his 1881 *Epilepsy and Other Chronic Convulsive Diseases*.)[2]

Gowers was especially drawn to the visual symptoms of epilepsy (he had previously written a book on ophthalmology), and he enjoyed describing simple visual seizures, as with one patient, for whom, he wrote:

> The warning was always a blue star, which appeared to be opposite the left eye, and to come nearer until consciousness was lost. Another patient always saw an object, not described as light, before the left eye, whirling round and round. It seemed to come nearer and nearer, describing larger circles as it approached, until consciousness was lost.

Jen W., an articulate young woman, came to see me several years ago. She told me that when she was four, she saw "a ball of colored lights on the right side, spinning, very defined." The ball of colored lights spun for a few seconds and was succeeded by a greyish cloud to the right, obscuring her vision to that side for two or three minutes.

She had further visions of the spinning ball, always in the same place, four or five times a year, but she assumed that this

---

2. Beginning in 1861, when he was twenty-four, Hughlings Jackson published many major papers—on epilepsy, aphasia, and other subjects, as well as what he called "evolution and dissolution in the nervous system." A selection of these, filling two large volumes, was published in 1931, twenty years after his death. In his later years, Jackson published a series of twenty-one short, gemlike papers in the *Lancet* under the title *Neurological Fragments*. These were collected and published in book form in 1925.

was normal, something everyone saw. When she was six or seven, the attacks took on a new aspect: the colored ball was followed by a headache on one side of her head, often accompanied by an intolerance of light and sound. She was taken to a neurologist, but an EEG and CAT scan revealed nothing, and Jen was diagnosed with migraine.

When she was thirteen or so, the attacks became longer, more frequent, and more complicated. Sometimes these frightening attacks led to complete blindness for several minutes, along with an inability to understand what people were saying. When she tried to talk, she could only utter gibberish. At this point, she was diagnosed with "complicated migraine."

When she was fifteen, Jen had a grand mal seizure—she had a convulsion and fell to the floor, unconscious. She had many EEGs and an MRI, all of which were interpreted as normal, but finally a detailed investigation by an epilepsy specialist revealed a clear epileptic focus in the left occipital lobe and an area of abnormal cortical architecture in the same area. She was put on antiepileptic drugs, and these prevented further convulsions but did little to help with her purely visual seizures, which became increasingly frequent, sometimes occurring many times a day. She said they could be precipitated by "bright sunlight, flickering shadows, or brightly colored scenes with movement and fluorescent lights." This extreme sensitivity to light drove her to a very restricted life, an effectively nocturnal and crepuscular existence.

Since her visual seizures did not respond to medication, a surgical approach was suggested, and when Jen was twenty, she had the abnormal area in her left occipital lobe removed. Before the surgery, while the occipitotemporal cortex was being mapped by electrical stimulation, she saw "Tinkerbell" and "cartoon figures." This was the only time she has ever had complex visual hallucinations; her visual seizures are normally of a simple sort, with the spinning ball to the right or, occasionally, a shower of "sparklers" in this area.

The immediate effect of the surgery was very good. She was thrilled that she no longer had to stay inside, and she went back to teaching gymnastics. She found that a very small dose of antiepileptic medication could now control most of her visual seizures, although she remained sensitive to stress, missing meals, not getting enough sleep, and flickering or fluorescent lights. Her surgery left her with blindness in the lower right quadrant of her visual field, and although she can navigate pretty well in the world with this blind spot, she avoids driving. Her symptoms returned, though less severely, a few years after the surgery. She says, "Epilepsy is a major challenge in my life, but I've developed strategies to manage it." She is working now on a PhD in biomedical engineering (with a focus on neuroscience), not least because of the intricate ways in which a neurological disorder has affected her own life.

When the epileptic focus lies at higher levels of the sensory cortex, in the parietal or the temporal lobes, the epileptic hallucinations may be much more complex. Valerie L., a gifted twenty-eight-year-old doctor, had what were called "migraines" from an early age—one-sided headaches preceded by twinkling blue dots. But when she was fifteen, she had a new, unprecedented experience. She said, "I had run a ten-mile race the day before . . . the next day I felt very strange. . . . I had a six-hour nap after a full night of sleep, which was most atypical for me, and then I went to temple with my family: it was a long service, a lot of standing." She started to see halos around objects and said to her sister, "Something weird is happening." And then a glass of water at which she was looking suddenly "multiplied itself," so that she saw glasses of water wherever she looked, dozens of them, covering the walls and the ceiling. This went on for perhaps five seconds, "the longest five seconds of my life," she said.

Then she lost consciousness. She came to in an ambulance, hearing the driver say, "I have a fifteen-year-old girl with a seizure," and then realized with a start that *she* was the girl.

When she was sixteen, she had a second, similar attack and was put on antiepileptic medication for the first time.

A third grand mal seizure occurred a year later. Valerie saw vague black shapes in the air ("like Rorschach ink blots"), and as she continued to look, these transformed themselves into faces—her mother's face and the faces of other relatives. The faces were motionless, flat, two-dimensional, and "like negatives," so that light-skinned faces were seen as dark, and vice versa. They had wavering edges, "as if enveloped in flame," in the thirty seconds before she had a convulsion and lost consciousness. After this, her doctors changed her antiepileptic medication, and she has had no more grand mal seizures since, though she continues to get visual auras or visual seizures, on average twice a month, more if she is stressed or sleep-deprived.

On one occasion, when Valerie was in college, she felt weak and not quite herself, so she went to her parents' house for the evening. She and her mother were sitting and talking as Valerie lay in bed, when she suddenly "saw" e-mails she had received earlier in the day plastered all over her bedroom. One particular e-mail was multiplied, and one of its images was superimposed on her mother's face, although she could see the face through it. The image of the e-mail was so clear and exact that she could read every word. Objects from her dorm room appeared everywhere she looked. It was a particular object, whether perceived or remembered, that got multiplied, never a whole scene. Her visual multiplications and reiterations are now of familiar faces for the most part, "projected" onto the walls, the ceiling, any available surface. This sort of spreading of visual perceptions in space (polyopia) and in time (palinopsia) was vividly described by Macdonald Critchley, who first used the term palinopsia (he originally called it paliopsia).

Valerie may also experience perceptual changes in relation to her seizures; indeed, her first intimation of a seizure is sometimes that her own reflection looks different—her eyes, in particular. She may feel, "This is not me," or "It's a close relative." If she can go to sleep, she can avert a seizure. But if she has not been able to sleep well, other people's faces may also look different the next morning—"strange" and distorted, especially around the eyes, though not so much that they are unrecognizable. Between attacks, she may have the opposite feeling—a hyperfamiliarity, so that everyone seems familiar to her. It is a feeling so overwhelming that sometimes she cannot resist greeting a stranger, even though, intellectually, she can say to herself, "This is just an illusion. It seems most unlikely that I have ever met this person."

Despite her epileptic auras, Valerie lives a full and productive life, keeping up with a demanding career. She is reassured by three things: that she has not had a generalized seizure for ten years, that whatever is provoking her attacks is not progressive (she had a minor head injury when she was twelve and probably has a small temporal lobe scar from that injury), and that medication can provide adequate control.

Both Jen and Valerie were initially misdiagnosed as having "migraine"—such confusion of epilepsy and migraine is not uncommon. Gowers was at pains to differentiate them in his 1907 book, *The Border-land of Epilepsy*, and his lucid descriptions bring out some of the differences between the two ailments as well as some of the similarities. Both migraine and epilepsy are paroxysmal—they present themselves suddenly, go through their course, and then disappear. Both show a slow movement or "march" of symptoms and the electrical disturbance underlying them—in migraine this takes fifteen or twenty minutes; in epilepsy it is often just a matter of seconds. It is unusual for people with migraine to have complex hallucinations, whereas epilepsy commonly affects higher parts of the brain; there it may evoke very complex, multisensorial "reminiscences" or dreamlike fantasies, like one of Gowers's patients, who saw "London in ruins, herself the sole spectator of this desolate scene."

Laura M., a psychology major in college, at first ignored her "strange attacks" but finally consulted an epilepsy specialist, who found that she was "experiencing stereotypic episodes of déjà vu, visual and emotional flashbacks of a dream or series of dreams, usually one of five dreams . . . which she had in the past ten years." These could happen several times daily and were aggravated by tiredness or by marijuana. When she started taking an antiepileptic medication, her attacks decreased in severity and frequency, but she had increasingly unacceptable side effects—in particular, a feeling of overstimulation followed by a "crash" later in the day. She took herself off the medication and reduced her use of marijuana, and now her attacks are at a tolerable level, perhaps half a dozen per month. They last only a few seconds, and although the internal feeling is overwhelming and she may "zone out" a little, others might not notice anything amiss. The only physical symptom she feels during these attacks is an impulse to roll her eyeballs back, which she resists when others are around.

When I met Laura, she said she had always had vivid, richly colored dreams that she could easily remember, and she characterized most of them as "geographic," involving complex landscapes. She felt that the visual hallucinations or flashbacks she had in her seizures all drew on the landscapes of those dreams.

One such dreamscape was Chicago, where she lived as a teenager. Most of her seizures transported her to this dream Chicago—she has drawn maps of it, which contain actual landmarks, but in which the topography is strangely transformed. Other dreamscapes center around the hill in another city, where her university is situated. "For a few seconds," she told me, "I flash back to a dream I have had, into the world of that dream, being in a different time and place. The places are 'familiar,' but don't really exist."

Another dreamscape, often reexperienced in seizures, is a transformed version of a hill town in Italy where she lived for a while. There is another, frightening one: "I'm with my little sister, on some sort of beach. We're being bombed. And

I lose her. . . . People are being killed." Sometimes, she says, the dreamscapes blend together, a hill somehow turning into a beach. There are always strong emotional components—fear or excitement, usually—and these emotions can dominate her for fifteen minutes or so after the actual attack.

Laura has quite a lot of apprehension about these odd episodes. On one of her maps, she wrote, "This all really scares me. Please, help me any way possible. Thanks!" She says she would give a million dollars to be free of these attacks—but she also feels they are a portal to another form of consciousness, another time and place, another world, although that portal is not under her control.

In his 1881 *Epilepsy,* Gowers gave many examples of simple sensory seizures and noted that auditory warnings of a seizure were as common as visual ones. Some of his patients spoke of hearing "the sound of a drum," "hissing," "ringing," "rustling," and sometimes more complex auditory hallucinations, such as music. (Music can be a hallucination in seizures, but real music may also trigger seizures. In *Musicophilia,* I described several examples of such musicogenic epilepsy.)[3]

There may also be chewing and lip-smacking movements in a complex partial seizure, occasionally accompanied by hal-

---

3. David Ferrier, a contemporary of Gowers's, moved to London in 1870, where he was mentored by Hughlings Jackson (Ferrier became a great experimental neurologist in his own right—he was the first to use electrical stimulation to map the monkey's brain). One of Ferrier's epileptic patients had a remarkable synesthetic aura, in which she would experience "a smell like that of green thunder." (This is quoted by Macdonald Critchley in his 1939 paper on visual and auditory hallucinations.)

lucinatory tastes.[4] Olfactory hallucinations, either alone as an isolated aura or as a part of complex seizure, may occur in various forms, as David Daly described in a 1958 review paper. Many of these hallucinatory smells seem unidentifiable or indescribable (except as "pleasant" or "unpleasant"), even though a patient will have the same smell in every seizure. One of Daly's patients said his hallucinatory smell odor was "somewhat like the smell of frying meat"; another said it was "like passing a perfume shop." One woman would experience an odor of peaches so vivid, so real, that she was certain there must be peaches in the room.[5] Another patient had a "reminiscence" associated with hallucinatory smells which "seemed to recall odors in his mother's kitchen when he was a child."

In 1956, Robert Efron, a naval physician, provided an extraordinarily detailed description of his patient Thelma B., a middle-aged professional singer. Mrs. B. experienced olfactory symptoms in her seizures, and she also gave a striking description of what Hughlings Jackson called doubled consciousness:

4. Hughlings Jackson described such seizures in 1875 and thought they might originate from a structure in the brain located beneath the olfactory cortex, the uncinate gyrus. In 1898 Jackson and W. S. Colman were able to confirm this by autopsy in Dr. Z., a patient who had died of an overdose of chloral hydrate. (More recently, David C. Taylor and Susan M. Marsh have recounted the fascinating history of Dr. Z., an eminent physician named Arthur Thomas Myers whose brother, F. W. H. Myers, had founded the Society for Psychical Research.)

5. In the 1946 film *A Matter of Life and Death* (called *Stairway to Heaven* in the United States), David Niven's character, a pilot, has complex epileptic visions that are always preceded by an olfactory hallucination (the smell of burnt onions) and a musical one (a recurrent theme of six notes). Diane Friedman has written a fascinating book about this, indicating how meticulous the director, Michael Powell, was in consulting neurologists about the forms of epileptic hallucinations.

I can be perfectly well in every way when suddenly I feel snatched away. I seem to feel as if I'm in two places at once but in neither place at all—it is a feeling of being remote. I can read, write and talk and can even sing my lyrics. I know exactly what is going on but I somehow don't seem to be in my own skin. . . . When this feeling happens I know that I'm going to have a convulsion. I keep trying to stop it from happening. No matter what I do, it always comes. Everything goes ahead like a railroad schedule. At this part of my attack I feel very active. If I'm home I make beds, dust, sweep or do the dishes. My sister says that I do everything at breakneck speed—I rush around like a chicken with his head cut off. But to me it all seems to be in slow motion. I am very interested in the time, I'm always looking at my watch and asking someone the time every few minutes. That is why I know exactly how long this part of the attack lasts. It has been as short as ten minutes or may last the better part of a day; it is real hell then. Usually it lasts about twenty to thirty minutes. All this time I feel that I'm remote. It is like being outside a room and looking in through a keyhole, or as if I'm God just looking down on the world but not belonging to it.

At about the halfway point in her seizure, Mrs. B. said, she would get a "funny idea" in her head involving the anticipation of a smell:

I expect to smell something at any moment, but I don't yet. . . . The first time it ever happened, I was out in the country and I was feeling funny. I was in a field picking forget-me-nots. I remember very well that I kept smelling these flowers even though I knew they had no odour. For about half an hour I kept sniffing them because I was sure they would begin to smell soon. . . even though I knew perfectly well at that time that forget-me-nots have no odour at all. . . . I know it and don't know it at the same time.

In this second phase of her epileptic aura, Mrs. B. continued to feel more and more "remote," until finally she knew a convulsion was near. She would lie on the floor, away from the furniture, to avoid hurting herself during the convulsion. Then, she said:

Just when I seem to be as remote as I possibly could get, I suddenly get a smell like an explosion or a crash. There is no buildup. It is all there at once. At the same moment that the smell crashes through, I'm back in the real world—I no longer feel remote. The smell is a disgusting sweet, penetrating odour like very cheap perfume. . . . Everything seems very quiet. I don't know if I can hear. I am all alone with the smell.

The smell would last for a few seconds and then go away, though the silence remained for five or ten seconds, until she heard a voice off to her right calling her name. She said:

This is not like hearing a voice in a dream. It is a *real* voice. Every time I hear it I fall for it. It is not a man's voice or a woman's voice. I don't recognize it. There is one thing that I do know and that is if I turn towards the voice I have a convulsion.

She would try hard not to turn towards the voice, but it was irresistible. Finally, she would lose consciousness and have a convulsion.

Gowers had a "favorite" seizure, one that he returned to in his writing many times, for this patient, like Thelma B., had an epileptic aura that involved many different sorts of hallucinations, unfolding in a "march" or stereotyped progression of symptoms. This showed Gowers how an epileptic excitation might move about the brain, stimulating first one part, then another, and evoking corresponding hallucinations as it did so. He first described this patient in his 1881 book *Epilepsy*:

> The patient was an intelligent man, twenty-six years of age, and all his attacks began in the same manner. First there was a sensation [under the ribs, on the left side] "like pain with a cramp;" then, this sensation continuing, a kind of lump seemed to pass up the left side of the chest, with a "thump, thump," and when it reached the upper part of the chest it became a "knocking," which was heard as well as felt. The sensation rose up to the left ear, and then was like the "hissing of a railway engine," and this seemed to "work over his head." Then he suddenly and invariably saw before him an old woman in a brown-stuff dress, who offered him something which had the smell of Tonquin beans. The old woman then disappeared, and two great lights came before him—round lights, side by side, which got nearer and nearer with a jerking motion. When the lights appeared the hissing noise ceased, and he felt a choking sensation in the throat, and lost consciousness in the fit, which, from the description, was undoubtedly epileptic.

For most people, focal seizures always consist of the same symptoms repeated with little or no variation, but others may have a large repertoire of auras. Amy Tan, the novelist, whose epilepsy may have been caused by Lyme disease, described her hallucinations to me.

"When I realized the hallucinations were seizures," she said, "I found them fascinating as brain quirks. I tried to notice the details of the ones that repeated." And, being a writer, she gave all of her repeating hallucinations names. The most frequent one she calls the "Illuminated Spinning Odometer." She describes it as

> what you might see on the dash of your car at night . . . except the numbers begin spinning more and more rapidly, like a gas pump giving you a running tally of the cost of gas. After about twenty seconds, the numbers begin to disintegrate and the odometer itself falls apart, and gradually disappears. Because it happened so often . . . I made it a game to see if I could name the numbers as they were falling, or to see if I could control the speed of the odometer or make the hallucination last longer. I could not.

None of her other hallucinations moved. For a time, she would often see

> the figure of a woman in long white Victorian dress in the foreground of a scene with other people in the background. It looked like a faint Victorian photograph, or a black and white version of one of those Renoir paintings of people in the park. . . . The figure was not looking at me, not moving. . . . I did not mistake it as a live scene or real people. The image had no significance to anything in my life. I did not feel any heightened emotions associated with it.

She sometimes has unpleasant odor hallucinations or physical sensations, "the ground beneath me wobbling, for instance," she says, adding, "I have to ask others if an earthquake is happening."

She often experiences déjà vu but finds her occasional jamais vu much more disturbing:

> The first time it happened, I remember looking at a building I had passed hundreds of times and thinking I had never noticed it was that color or shape, etc. And I then looked at everything around me, and none of it looked familiar. It was so disorienting I could not move an inch further. In the same way, I would sometimes not *recognize* my home, but I knew I was *in* my home. I had learned to be patient and wait for it to pass in twenty or thirty seconds.

Amy remarks that her seizures most often occur when she is waking up or dozing off. She occasionally sees "Hollywood aliens" dangling from the ceiling. They look like "someone's inept attempt to make an alien creature for a movie set . . . like a spider with a Darth Vader–like helmet head."

She emphasizes that the images have no personal relevance, are not related to anything that happened that day, and carry no special associations or emotional significance. "They do not stay in my mind as anything to think about," she observes. "They are more like the detritus of those parts of dreams that mean nothing, like random images arbitrarily flashed in front of me."

Stephen L., an affable, outgoing man, first consulted me in the summer of 2007. He brought with him his "neurohistory," as he called it—seventeen pages of single-spaced typing—adding that he had "a little graphomania." He said his problems started after an accident thirty years before, when his car was broadsided by another, and his head slammed against the windshield. He suffered a severe concussion but seemed to recover fully after a few days. Two months later, he started to have brief attacks of déjà vu: he would suddenly feel that whatever he was experiencing, doing, thinking, or feeling he had already experienced, done, thought, or felt before. At first he was intrigued by these brief convictions of familiarity and found them pleasant ("like the breeze going past my face"), but within a few weeks he was getting them thirty or forty times a day. On one occasion, to prove that the feeling of familiarity was an illusion, he stamped his foot, threw one leg high in the air, and did a sort of Highland fling in front of a washroom mirror. He *knew* he had never done such a thing before, but it *felt* as though he were repeating something he had done many times.

His attacks became not only more frequent but more complex, the déjà vu being only the start of a "cascade" (as he put it) of other experiences, which, once started, would move forward irresistibly. The déjà vu would be followed by a sharp icy or burning pain in the chest, then by an alteration of hearing, so that sounds become louder, more resonant, seemed to reverberate all around him. He might hear a song as clearly as if it were being sung in the next room, and what he heard would always be a specific performance of the song—for example, a particular Neil Young song ("After the Gold Rush") exactly as he had heard it during a concert at his college the year before. He might then go on to experience "a bland, pungent smell" and a taste "which corresponded with the smell."

On one occasion Stephen dreamt he was having one of his aura cascades and woke to find that he was indeed in the midst of one. But then to the usual cascade was added a strange out-of-body experience, in which he seemed to be looking down at his body as it lay in bed, through an elevated open window. This out-of-body experience seemed real—and very frightening. Frightening, in part, because it suggested to him that more and more of his brain was being involved in his seizures, and that things were getting out of control.

Nonetheless, he kept these attacks to himself until Christmas of 1976, when he had a convulsion, a grand mal seizure; he was in bed with a girl at the time, and she described it to him. He consulted a neurologist, who confirmed that he had temporal lobe epilepsy, probably caused by injury to the right temporal lobe sustained during the car crash. He was put on antiepileptics—first one, then others—but he continued to have temporal lobe seizures almost daily and two or more grand mal seizures a month. Finally, after thirteen years of trying different antiepileptic medications, Stephen consulted another neurologist for evaluation and consideration of possible surgery.

In 1990, Stephen had surgery to remove an epileptic focus in his right temporal lobe, and he felt so much better after the surgery that he decided to wean himself off medication. Then, unfortunately, he had another car accident, after which his seizures returned. These were not responsive to medication, and he had to have much more extensive brain surgery in 1997. Nevertheless, he continues to need antiepileptic medication and to have various seizure symptoms.

Stephen feels that there has been a "metamorphosis" in his personality since the start of his seizures, that he has become "more spiritual, more creative, more artistic"—specifically, he wonders whether "the right side" of his brain (as he puts it) is being stimulated, coming to dominate him. In particular, music has assumed greater and greater importance for him. He had taken up the harmonica in his college days, and now, in his fifties, he plays "obsessively," for hours. He often writes or draws for hours at a time, too. He feels that his personality has become "all or none"—he may be either hyperfocused or completely distracted. He has also developed a tendency to sudden rage: on one occasion, when a car cut him off, he attacked the offender physically, hurling a can at his car, then punching him. (He wonders, in retrospect, whether some seizure activity played a part in this.) Despite all his problems, Stephen L. is able to continue working in medical research, and he remains an engaging, sensitive, and creative person.

There was little that Gowers or his contemporaries could do for patients with complex or focal seizures, other than giving them sedative drugs like bromides. Many patients with epilepsy, especially temporal lobe epilepsy, were considered to be "medically intractable" until the introduction of the first specific antiepileptic drug in the 1930s—and even then the most severely affected patients could not be helped. But the 1930s also saw a much more radical, surgical approach, undertaken by Wilder Penfield, a brilliant young American neurosurgeon working in Montreal, and his colleague Herbert Jasper. In order to remove the epileptic focus in the cerebral cortex, Penfield and Jasper first had to find it by mapping the patient's temporal lobe, and this required the patient to be fully conscious. (Local anesthesia is used when opening the skull, but the brain itself is insensitive to touch and pain.) Over a twenty-year period, the "Montreal procedure" was carried out in more than five hundred patients with temporal lobe epilepsy. These people had very diverse seizure symptoms, but forty or so of them had what Penfield termed "experiential seizures," in which, seemingly, a fixed and vivid memory of the past would suddenly burst into the mind with hallucinatory force, causing a doubling of consciousness: a patient would feel equally that he was in the operating room in Montreal and that he was, say, riding horseback in a forest. By systematically going over the surface of the exposed temporal cortex with his electrodes, Penfield was able to find particular cortical points in each patient where stimulation caused a sudden, involuntary recall—an experiential seizure.[6] Removal of these points could prevent further such seizures, without affecting the memory itself.

---

6. Penfield was a great physiologist as well as a neurosurgeon, and in the process of searching for epileptic foci, he was able to map most of the basic functions of the living human brain. He showed, for example, exactly where sensations and movements of specific body parts were represented in the cerebral cortex—his sensory and motor homunculi are iconic. Like Weir Mitchell, Penfield was an engaging writer, and after he and Herbert Jasper published their magnum opus, *Epilepsy and the Functional Anatomy of the Human Brain*, in 1958, he continued to write about the brain, as well as writing novels and biographies, until his death at eighty-six.

Penfield described many examples of experiential seizures:

> At operation it is usually quite clear that the evoked experiential response is a random reproduction of whatever composed the stream of consciousness during some interval of the patient's past life. . . . It may have been a time of listening to music, a time of looking in at the door of a dance hall, a time of imagining the action of robbers from a comic strip . . . a time of lying in the delivery room at birth, a time of being frightened by a menacing man, a time of watching people enter the room with snow on their clothes. . . . It may have been a time of standing on the corner of Jacob and Washington, South Bend, Indiana.

Penfield's notion of actual memories or experiences being reactivated has been disputed. We now know that memories are not fixed or frozen, like Proust's jars of preserves in a larder, but are transformed, disassembled, reassembled, and recategorized with every act of recollection.[7]

---

7. For Gowers and his contemporaries in the early twentieth century, memories were imprints in the brain (as for Socrates they were analogous to impressions made in soft wax)—imprints that could be activated by the act of recollection. It was not until the crucial studies of Frederic Bartlett at Cambridge in the 1920s and 1930s that this classical view could be disputed. Whereas Ebbinghaus and other early investigators had studied rote memory—how many digits could be remembered, for instance—Bartlett presented his subjects with pictures or stories and then questioned and requestioned them over a period of months. Their accounts of what they had seen or heard were somewhat different (and sometimes quite transformed) on each re-remembering. These experiments convinced Bartlett to think in terms not of a static thing called "memory," but rather a dynamic process of "remembering." He wrote:

> Remembering is not the re-excitation of innumerable fixed, lifeless and fragmentary traces. It is an imaginative reconstruction, or construction, built out of the relation of our attitude towards a whole active mass of organized past reactions or experience. . . . It is thus hardly ever really exact.

And yet, some memories do, seemingly, remain vivid, minutely detailed, and relatively fixed throughout life. This is especially so with traumatic memories or memories carrying an intense emotional charge and significance. Penfield was at pains, however, to emphasize that epileptic flashbacks seem to lack any such special qualities.[8] "It would be very difficult to imagine," he wrote, "that some of the trivial incidents and songs recalled during stimulation or epileptic discharge could have any possible emotional significance to the patient, even if one is acutely aware of this possibility." He felt that the flashbacks consisted of "random" segments of experience, fortuitously associated with a seizure focus.

Curiously, though Penfield described such a variety of experiential hallucinations, he made no reference to what we now call "ecstatic" seizures—seizures that produce feelings of ecstasy or transcendent joy, such as Dostoevsky described. Dostoevsky's seizures started in childhood, but they became frequent only in his forties, after his return from exile in Siberia. In his occasional grand mal attacks, he would emit (his wife wrote) "a fearful cry, a cry that had nothing human about it," and then fall to the floor, unconscious. Many of these attacks were preceded by a remarkable mystical or ecstatic aura—but sometimes there would be only the aura, without any subse-

---

8. Penfield sometimes used the term "flashback" for experiential hallucinations. The term is also used in quite different contexts, as in post-traumatic flashbacks, where there are recurrent hallucinatory replayings of traumatic events.

The term "flashback" is also used for a sudden, transient reexperiencing of a drug effect—suddenly feeling, for example, the effects of LSD, even though one has not taken it for months.

quent convulsions or lack of consciousness. The first occurred one Easter Eve, as his friend Sophia Kowalewski wrote in her *Childhood Recollections* (Alajouanine quotes this in his paper on Dostoevsky's epilepsy). Dostoevsky was talking with two friends about religion when a bell started to toll midnight. Suddenly he exclaimed, "God exists, He exists!" He later went into detail about the experience:

> The air was filled with a big noise and I tried to move. I felt the heaven was going down upon the earth and that it had engulfed me. I have really touched God. He came into me myself, yes God exists, I cried, and I don't remember anything else. You all, healthy people, he said, can't imagine the happiness which we epileptics feel during the second or so before our fit. . . . I don't know if this felicity lasts for seconds, hours or months, but believe me, for all the joys that life may bring, I would not exchange this one.

He gave similar descriptions on a number of other occasions, and endowed several of the characters in his novels with seizures akin to, and sometimes identical with, his own. One such involves Prince Myshkin in *The Idiot*:

> During these moments as rapid as lightning, the impression of the life and the consciousness were in himself ten times more intense. His spirit and his heart were illuminated by an immense sense of light; all his emotions, all his doubts, all his anxiety calmed together to be changed into a sovereign serenity made up of lighted joy, harmony and hope; then, his reason was raised up to the understanding of the final cause.

There are also descriptions of ecstatic seizures in *The Devils*, *The Brothers Karamazov*, and *The Insulted and the Injured*, while in *The Double* there are descriptions of "forced thinking" and "dreamy states" almost identical with what Hughlings Jackson was describing at much the same time in his great neurological articles.

Over and above his ecstatic auras—which always seemed to Dostoevsky revelations of ultimate truth, direct and valid knowledge of God—there were remarkable and progressive changes in his personality throughout the later parts of his life, his time of greatest creativity. Théophile Alajouanine, a French neurologist, observed that these changes were clear when one compared Dostoevsky's early, realistic works with the great, mystical novels he wrote in later life. Alajouanine suggested that "epilepsy had created in the person of Dosto-evsky a 'double man' . . . a rationalist and a mystic; each hav-ing the better of the other according to the moment . . . [and] more and more the mystical one seems to have prevailed."

It was this change, seemingly progressing even between Dostoevsky's seizures ("interictally," in neurological jargon), that especially fascinated the American neurologist Norman Geschwind, who wrote a number of papers on the subject in the 1970s and 1980s. He noted Dostoevsky's increasingly obsessive preoccupation with morality and proper behavior, his growing tendency to "get embroiled in petty arguments," his lack of humor, his relative indifference to sexuality, and, despite his high moral tone and seriousness, "a readiness to become angry on slight provocation." Geschwind spoke of all this as an "interictal personality syndrome" (it is now called "Geschwind syndrome"). Patients with it often develop an intense preoccupation with religion (Geschwind referred to this as "hyper-religiosity"). They may also develop, like Stephen L., compulsive writing or unusually intense artistic or musical passions.

Whether or not an interictal personality syndrome develops—and it does not seem to be universal or inevitable in those who have temporal lobe epilepsy—there is no doubt that those who have ecstatic seizures may be profoundly moved by them, and even actively seek to have more such seizures. In 2003, Hansen Asheim and Eylert Brodtkorb, in Norway, published a study of eleven patients with ecstatic seizures; eight of them wished to experience their seizures again, and of these, five found ways to induce them. More than any other sort of seizure, ecstatic seizures may be felt as epiphanies or revelations of a deeper reality.

Orrin Devinsky, a former student of Geschwind's, has been a pioneer himself in the investigation of temporal lobe epilepsy and the great range of neuropsychiatric experiences which may be associated with it—autoscopy, out-of-body experiences, déjà vu and jamais vu, hyperfamiliarity, and ecstatic states during seizures, as well as personality changes between seizures. He and his colleagues have been able to perform clinical and video EEG monitoring in patients as they are having ecstatic-religious seizures, and thus to observe the precise coinciding of their "theophanies" with seizure activity in temporal lobe seizure foci (nearly always these are right-sided).[9]

---

9. One such patient, who had very little in the way of religious interests as an adult, had his first religious seizure at a picnic, as Devinsky described to me: "His friends observed at first that he stared, became pale, and was unresponsive. Then suddenly, he began to run in circles for two or three minutes yelling, 'I am free! I am free! . . . I am Jesus! I am Jesus!' "

Such revelations may take different forms; Devinsky has told me of one woman who, following a head injury, started to have brief episodes of déjà vu and a strange, indescribable smell. After a cluster of these complex partial seizures, she entered an exalted state in which God, with the form and voice of an angel, told her to run for Congress. Though she had never been religious or political before, she acted on God's words at once.[10]

On occasion, ecstatic hallucinations can be dangerous, although this is very rare. Devinsky and his colleague George Lai described how one of their patients had a seizure-related vision in which "he saw Christ and heard a voice that commanded him to kill his wife and then himself. He proceeded to act upon the hallucinations," killing his wife and then stabbing himself. This patient ceased to have seizures after the seizure focus in his right temporal lobe was removed.

Such epileptic hallucinations bear a considerable resemblance to the command hallucinations of psychosis, even

---

The patient later had a similar seizure which was recorded on video EEG, and, just before the seizure, Devinsky noted, the patient was slow to respond and disoriented regarding time and place:

> When asked if there was anything wrong, he replied: "nothing is wrong, I am doing well . . . I am very happy" and when asked whether he knew where he was, he replied with a smile and a surprised look: "Of course I know. I am in heaven right now; . . . I am fine. "

He remained in this state for ten minutes, then went on to a generalized seizure. Later, he remembered his ecstatic aura "as if it were a vivid and happy dream" from which he had now awoken, but he had no memory of the questions put to him during the aura.

10. She ran as a Republican in a district that had been Democratic for a very long time, and lost by only a narrow margin. Whenever she appeared in public during her campaign, she said that God had told her to run, and this apparently persuaded thousands of people to vote for her, despite her manifest lack of political experience or skills.

though the epileptic patient may have no psychiatric history. It takes a strong (and skeptical) person to resist such hallucinations and to refuse them either credence or obedience, especially if they have a revelatory or epiphanic quality and seem to point to a special—and perhaps exalted—destiny.

As William James observed, an acute and passionate religious conviction in a single person can sway thousands of people. The life of Joan of Arc exemplified this. People have puzzled for nearly six hundred years as to how a farmer's daughter with no formal education could have found such a sense of mission and succeeded in getting thousands of others to aid her in an attempt to drive the English out of France. The early hypotheses of divine (or diabolic) inspiration have given way to medical ones, with psychiatric diagnoses vying with neurological ones. Much evidence is available from the transcripts of her trial (and her "rehabilitation" twenty-five years later) and from the recollections of contemporaries. None of these is conclusive, but they do suggest, at least, that Joan of Arc may have had temporal lobe epilepsy with ecstatic auras.

Joan experienced visions and voices from the age of thirteen. These came in discrete episodes lasting seconds or minutes at most. She was very frightened by the first visitation, but later she derived great joy and an explicit sense of mission from her visions. The episodes were sometimes precipitated by the sounds of church bells. Joan described her first "visitations":

I was thirteen when I had a Voice from God for my help and guidance. The first time that I heard this Voice, I was very much frightened; it was mid-day, in the summer, in my father's garden . . . I heard this Voice to my right, towards the Church; rarely do I hear it without its being accompanied also by a light. This light comes from the same side as the Voice. Generally it is a great light. . . . When I heard it for the third time, I recognized that it was the Voice of an Angel. This voice has always guarded me well, and I have always understood it; it instructed me to be good and to go often to Church; it told me it was necessary for me to come into France . . . it said to me two or three times a week: "You must go into France." . . . It said to me: "Go, raise the siege which is being made before the City of Orleans. Go!" . . . and I replied that I was but a poor girl, who knew nothing of riding or fighting. . . . There is never a day when I do not hear this Voice; and I have much need of it.

Many other aspects of Joan's putative seizures, as well as evidence of her clarity, her reasonableness, and her modesty, were explored in a 1991 article by the neurologists Elizabeth Foote-Smith and Lydia Bayne. While they present a very plausible case, other neurologists disagree, and one cannot hope to see the matter definitively resolved. The evidence is soft, as it must be for all historical cases.

Ecstatic or religious or mystical seizures occur in only a small number of those who have temporal lobe epilepsy. Is this because there is something special—a preexisting disposition to religion or metaphysical belief—in these particular people? Or is it because the seizure stimulates particular parts of the brain that serve to mediate religious feeling?[11] Both, of course,

---

11. The evidence here has been discussed in a number of books, including Kevin Nelson's *The Spiritual Doorway in the Brain: A Neurologist's Search for the God Experience*. It is also the theme of a novel, *Lying*

could be the case. And yet quite skeptical people, indifferent to religion, not given to religious belief, may—to their own astonishment—have a religious experience during a seizure.

Kenneth Dewhurst and A. W. Beard, in a 1970 paper, provided several examples of this. One related to a bus conductor who had an ecstatic seizure while collecting fares:

> He was suddenly overcome with a feeling of bliss. He felt he was literally in Heaven. He collected the fares correctly, telling his passengers at the same time how pleased he was to be in Heaven. . . . He remained in this state of exaltation, hearing divine and angelic voices, for two days. Afterwards he was able to recall these experiences and he continued to believe in their validity. . . . During the next two years, there was no change in his personality; he did not express any peculiar notions but remained religious. . . . Three years later, following three seizures on three successive days, he became elated again. He stated that his mind had "cleared." . . . During this episode he lost his faith.

He now no longer believed in heaven and hell, in an afterlife, or in the divinity of Christ. This second conversion—to atheism—carried the same excitement and revelatory quality as the original religious conversion. (Geschwind, in a 1974 lecture subsequently published in 2009, noted that patients with temporal lobe epilepsy might have multiple religious conver-

---

*Awake*, by Mark Salzman; the protagonist is a nun who has ecstatic seizures in which she communes with God. Her seizures, it turns out, are caused by a tumor in her temporal lobe, and it must be removed before it enlarges and kills her. But will its removal also remove her portal to heaven, preventing her from ever communing with God again?

sions and described one of his own patients as "a girl in her twenties who is now on her fifth religion."}

Ecstatic seizures shake one's foundations of belief, one's world picture, even if one has previously been wholly indifferent to any thought of the transcendent or supernatural. And the universality of fervent mystical and religious feelings—a sense of the holy—in every culture suggests that there may indeed be a biological basis for them; they may, like aesthetic feelings, be part of our human heritage. To speak of a biological basis and biological precursors of religious emotion—and even, as ecstatic seizures suggest, a very specific neural basis, in the temporal lobes and their connections—is only to speak of natural causes. It says nothing of the value, the meaning, the "function" of such emotions, or of the narratives and beliefs we may construct on their basis.

9

# Bisected:
# Hallucinations in the Half-Field

One does not see with the eyes; one sees with the brain, which has dozens of different systems for analyzing the input from the eyes. In the primary visual cortex, located in the occipital lobes, at the back of the brain, there are point-to-point mappings of the retina onto the cortex, and it is here that light, shape, orientation, and location in the visual field are represented. Impulses from the eyes take a circuitous route to the cerebral cortex, some of them crossing to the opposite side of the brain as they do so, so that the left half of the visual field of each eye goes to the right occipital cortex, and vice versa. If, therefore, one occipital lobe is damaged (as by a stroke, for example), there will be blindness or impaired vision in the opposite half of the visual field—a hemianopia.

Besides the impairment or loss of vision to one side, there may be positive symptoms, too—hallucinations in the blind or purblind area. About 10 percent of patients with sudden hemianopia get such hallucinations—and immediately recognize them to *be* hallucinations.

In contrast to the relatively brief and stereotyped hallucinations of migraine or epilepsy, the hallucinations of hemianopia may continue for days or weeks on end; and, far from being fixed or uniform in format, they tend to be ever changing. Here, one might envisage not a small knot of irritable cells discharging paroxysmally, as in an attack of migraine or epilepsy, but a large area of the brain—whole fields of neurons—in a state of chronic hyperactivity, out of control and misbehaving because of the lessening of forces that normally control or organize them. The mechanism here thus resembles that of Charles Bonnet syndrome.

While such notions were implicit in Hughlings Jackson's vision of the nervous system as having hierarchically ordered levels (the higher levels controlling the lower ones, and lower ones starting to behave independently, even anarchically, if released from control by damage at the higher levels), the idea of "release" hallucinations was made explicit by L. Jolyon West in his 1962 book *Hallucinations.* A decade later, David G. Cogan, an ophthalmologist, published an influential paper that included short, vivid histories of fifteen patients. Some of them had damage to their eyes, some had damage to their optic nerves or tracts, some had occipital lobe lesions, some had temporal lobe lesions, and some had lesions in the thalamus or the midbrain. Lesions in any of these different places, it seemed, could break the normal network of controls and lead to a release of complex visual hallucinations.

Ellen O. was a young woman who came to see me in 2006, about a year after surgery for a vascular malformation in her right occipital lobe. The procedure was a fairly simple one, sealing off the swollen vessels of the malformation. As her doctors had warned her, she had some visual problems following the procedure: a blurring of vision to the left side, as well as some agnosia and alexia—difficulties recognizing people and printed words (English words looked like "Dutch," she said). These difficulties prevented her from driving for six weeks and interfered with her reading and enjoyment of television, but they were transient. She also had visual seizures in the first weeks after surgery. These took the form of simple visual hallucinations, flashes of light and color to the left that lasted a few seconds. The seizures came several times a day at first but had practically ceased by the time she returned to work. She was not too concerned about them, for her doctors had warned her that she might experience such aftereffects.

What they had not warned her about was that she might develop complex hallucinations later. The first of these, about six weeks after her surgery, was of a huge flower, occupying most of the left half of her vision. This had been stimulated, she thought, by seeing an actual flower in bright, dazzling sunlight; it seemed to burn itself into her brain, and the vision of it persisted in the left half of her visual field, "like an afterimage"—but an afterimage that lasted not for a few seconds but for an entire week. The following weekend, after her brother visited, she saw his face—or, rather, part of his profile, just one eye and cheek—for several days.[1]

---

1. Before seeing Ellen O., I had never heard of visual perseveration of such duration. Visual perseveration of a few minutes may be associated with cerebral tumors of the parietal or temporal lobes or may occur in temporal lobe epilepsy. There are a number of such accounts in the medical literature, including one by Michael Swash, who described two people with temporal lobe epilepsy. One of them had attacks in which

Then she moved from abnormalities of perception—seeing things that were actually there, with perseveration or distortion—to hallucinations, seeing things that were not there. Visions of people's faces (including, at times, her own) became a frequent sort of hallucination. But the faces Ellen saw were "abnormal, grotesque, exaggerated," often just a profile with the teeth or perhaps one eye hugely magnified, completely out of scale with the rest of the features. At other times she saw figures with "simplified" faces, expressions, or postures—"like sketches or cartoons." Then Ellen started to hallucinate Kermit the Frog, the *Sesame Street* puppet, many times a day. "Why Kermit?" she asked. "He means nothing to me."

Most of Ellen's hallucinations were flat and still, like photographs or caricatures, though sometimes an expression would change. Kermit the Frog sometimes looked sad, sometimes happy, occasionally angry, though she could not connect his expressions with any of her own moods. Silent, motionless, ever changing, these hallucinations were almost continuous throughout her waking hours ("They are 24/7," she said). They did not occlude her vision but were superimposed like transparencies over the left half of her visual field. "They have been getting smaller lately," she told me. "Kermit the Frog is tiny

---

"his vision seemed to become fixed, so that an image was retained for several minutes. During these episodes the real world was seen through the retained image, which was clear at first, but then gradually faded."

Similar perseveration may occur with damage or surgery to an eye. My correspondent H.S. was blinded by a chemical explosion at the age of fifteen but had some sight restored by corneal surgery twenty years later. Following the operation, when his surgeon asked if he could now see the surgeon's hand, H.S. replied, "Yes"—but then was astonished to see the hand, or its image, preserving its exact shape and position, for several minutes afterward.

now. He used to occupy most of the left half, and now he's down to a little fraction of it." Ellen wondered whether she would have these hallucinations for the rest of her life. I said that I thought their diminution a very good sign; perhaps one day Kermit would be too small to see at all.

What was going on in her brain? she asked me. Why, above all, was she getting these odd and sometimes nightmarish hallucinations of grotesque faces? From what depths did they come? Surely it was not normal to imagine such things. Was she becoming psychotic, going mad?

I told her that the impairment of vision on one side following her surgery had probably led to heightened activity in parts of the brain higher up in the visual pathway, in the temporal lobes, where figures and faces are recognized, and perhaps in the parietal lobes, too; and that this heightened, at times uncontrolled, activity was causing her complex hallucinations and also the extraordinary persistence of vision, the palinopsia, she was experiencing. The particular hallucinations which so horrified her—of deformed and dismembered faces or faces with exaggerated, monstrous eyes or teeth—were, in fact, typical of abnormal activity in an area of the temporal lobes called the superior temporal sulcus. They were neurological faces, not psychotic ones.

Ellen wrote to me periodically with updates, and six years after our initial visit, she wrote: "I would not say that I am entirely recovered from my visual problems; more that I am living more harmoniously with them. My hallucinations are much smaller, but they are still there. Mostly I see the colorful orb all the time, but it no longer distracts me as much."

She still has some difficulty with reading, especially when she is tired. When she read a book recently, she said,

I lost a word or two in my color spot (I had a black/blind spot after surgery, but it turned into a colored spot a few weeks later, and I still have it. My hallucinations are around that spot.) . . . As I type now, after a very long day at work, there is a very faint black-and-white Mickey Mouse from the thirties just off center to the left. He's transparent, so I'm able to see my computer screen as I type. I do, however, make many mistakes typing, as I can't always see the key I need.

But Ellen's blind spot has not prevented her from pursuing graduate courses and even marathon running, as she reported with characteristic good humor:

I ran the New York City marathon in November and tripped on this metal ring, a piece of garbage, on the Verrazano Bridge a little before the second mile. It was on my left side, and I didn't even see it, as I was only looking to my right. I got back up and finished, although I did break a small bone in my hand—which, I think, makes for a wonderful running injury story. In the orthopedics waiting room when I was there, everyone else who had finished the marathon had knee or hamstring injuries.

While Ellen's complex hallucinations started several weeks after her operation, similar "release" hallucinations may appear almost immediately with sudden damage to the occipital cortex. This was the case with Marlene H., a woman in her fifties who came to see me in 1989. She told me that she had awoken one Friday morning in December 1988 with a headache and visual symptoms. She had had migraines for years, and at first she took this as just another visual migraine. But the visual symptoms were different this time: she saw "flashing lights all over . . . shimmering lights . . . arcs of lightning . . . like a Frankenstein thing," and these did not go away in a few minutes, like her usual migraine zigzags, but continued all through the weekend. Then, on Sunday evening, the visual disturbances took on a more complex character. In the upper part of the visual field, to the right, she saw a writhing form "like a Monarch caterpillar, black and yellow, its cilia glistening," along with "incandescent yellow lights, like a Broadway show, going up and down, on and off, nonstop." Though her doctor had reassured her that this was just "an atypical migraine," things went from bad to worse. On Wednesday, "the bathtub seemed to be crawling with ants . . . there were cobwebs covering the walls and ceiling . . . people seemed to have lattices on their faces." Two days later she started to experience gross perceptual disturbances: "My husband's legs looked really short, distorted, like someone in a trick mirror. It was funny." But it was less funny, and rather frightening, in the market that afternoon: "Everyone looked ugly, parts of their faces were gone, and eyes—there seemed a blackness in their eyes—everyone looked grotesque." Cars seemed to appear suddenly to the right. Testing her visual fields, waggling her fingers to either side, Marlene found that she could not see them on the right until they crossed the midline; she had lost all vision to the right side.

It was only at this point, days after her initial symptoms, that she was finally investigated medically. A CAT scan of her brain revealed a large hemorrhage in the left occipital lobe. There was little to be done therapeutically at this stage; one could only hope that there would be some resolution of her symptoms, some healing or adaptation with time.

After some weeks, the hallucinations and perceptual distortions, which had been largely confined to the right side, did start to die down, but Marlene was left with a variety of visual deficits. She could see, at least to one side, but was bewildered by what she saw: "I would have preferred to be blind," she told me, "instead of not being able to make sense of what I saw. . . . I had to go slowly, deliberately, to put things together. I would see my sofa, a chair—but I couldn't put it together. It did not add up, at first, to a 'scene.' . . . I was a very fast reader before. Now I was slow. The letters looked different."

"When she looks at her watch," her husband interpolated, "at first she can't process it."

Besides these problems of visual agnosia and visual alexia, Marlene was experiencing a sort of runaway visual imagery, outside her control. At one point, she saw a woman wearing a red dress on the street. Then, she said, "I closed my eyes. This woman, almost puppetlike, was moving around, took on a life of her own. . . . I realize that I had been 'taken over' by the image."

I kept in touch with Marlene at intervals and saw her most recently in 2008, twenty years after her stroke. She no longer had hallucinations, perceptual distortions, or runaway visual imagery. She was still hemianopic, but her remaining vision was good enough for her to travel independently and to work (which involved reading and writing, albeit at her own slow pace).

While Marlene experienced protracted perceptual changes as well as hallucinations after a massive occipital lobe hemorrhage, even a "little" occipital lobe stroke can evoke

striking, though transient, visual hallucinations. Such was
the case with a bright, deeply religious old lady whose hallucinations appeared, "evolved," then disappeared, all within
the space of a few days in July of 2008. I got a call from one of
the nurses in a nursing home where I work—we had worked
together for many years, and she knew that I was especially
interested in visual problems. She asked whether she could
bring her great-aunt Dot to see me, and between them, they
reconstructed the story. Aunt Dot told me that her vision had
seemed "blurry" on July 21, and the following day, "it was
like looking through a kaleidoscope . . . all this rotating color
going through," with sudden "lightning streaks" to the left.
She went to her doctor, who, finding that she had a hemianopia to the left, sent her to an emergency room. There it was
found that she had atrial fibrillation, and a CAT scan and MRI
showed a small area of damage in the right occipital lobe, probably the result of a blood clot dislodged by the fibrillation.

The following day, Aunt Dot saw "octagons with red centers . . . moving past me like a film strip, and the moving octagons changed into hexagonal snowflakes." On July 24, she saw
"an American flag, outstretched, as if flying."

On July 26, she saw green dots, like little balls, floating
to the left, and these turned into "elongated silvery leaves."
When her niece remarked that an early autumn was on the
way in Canada and the leaves were already changing color,
the hallucinated silvery leaves immediately turned reddish
brown. These ushered in a day full of complex visual hallucinations, including "daffodils in bouquets" and "fields of gold-
enrod." They were followed by a very particular image, which
was multiplied. When her niece visited that day, Aunt Dot
said, "I'm seeing sailor boys . . . one on top of the other, like

a film strip." They were colored, but flat and motionless and small, "like stickers." She did not recognize their origin until her niece reminded her that she (the niece) often used a sailor-boy sticker when she sent her aunt a letter—so here, the sailor boy was not a complete invention, but a reproduction of the stickers Aunt Dot had once seen, now multiplied.

The sailor boys were replaced by "fields of mushrooms" and then by a golden Star of David. A neurologist in the hospital had been wearing such a star prominently when he visited her, and she continued to "see" this for hours, though not multiplied like the sailor boys. The Star of David was superseded by "traffic lights, red and green, turning on and off," then by scores of tiny golden Christmas bells. The Christmas bells were replaced by a hallucination of praying hands. Then she saw "gulls, sand, waves, a beach scene," with the gulls flapping their wings. (Up to this point, apparently, there had not been movement within an image; she had seen only static images passing in front of her.) The flying gulls were replaced by "a Greek runner wearing a toga . . . he looked like an Olympic athlete." His legs were moving, as the gulls' wings had been. The next day she saw stacked and serried coat hangers—this was the last of her complex hallucinations. The day after that, she saw only lightning streaks to the left, as she had seen six days before. And this was the end of what she called her "visual odyssey."

Aunt Dot was not a nurse, like her great-niece, but she had worked for many years as a volunteer in the nursing home. She knew that she had had a small stroke on one side in the visual part of her brain. She realized that the hallucinations were caused by this and were probably transient; she did not fear that she was losing her mind. She did not for a moment think that her hallucinations were "real," although she observed that they were quite unlike her normal visual imagery—much more detailed, more brightly colored, and, for the most part, independent of her thoughts or feelings. She was curious and intrigued, and so she made a careful note of the hallucinations as they occurred and tried to draw them. Both she and her niece wondered why particular images popped up in her hallucinations, to what extent they reflected her life experiences, and how much they might have been prompted by her immediate environment.

She was struck by the sequence of her hallucinations—that they had gone from simple and unformed to more complex, and then back to simple before disappearing. "It's like they moved up the brain, then down again," she said. She was struck by how things she had seen could change into similar forms: octagons turning into snowflakes, blobs into leaves, and perhaps gulls into Olympic athletes. She observed that, in two instances, she had hallucinated something she had seen shortly before: the neurologist's Star of David and the sailor-boy stickers. She noted a tendency to "multiplication"—bunches of daffodils, fields of flowers, octagons galore, snowflakes, leaves, gulls, scores of Christmas bells, and multiple copies of the sailor-boy stickers. She wondered whether the fact that she was a deeply religious Catholic who prayed several times a day had played a part in her seeing a hallucination of praying hands. She was struck by the way in which the silvery leaves she was seeing instantly turned reddish brown when her niece said, "The leaves are changing." She thought the Olympic runner might have been provoked by the fact that the 2008 Olympic Games were coming up, with constant previews on television. I found it impressive and moving that this old lady, curious and intelligent, though not intellectual, would observe her own hallucinations so calmly and thoughtfully and, without being prompted, raise virtually all the questions a neurologist might ask about them.

If one loses half the visual field from a stroke or other injury, one may or may not be aware of the loss. Monroe Cole, a neurologist, became aware of his own field loss only by doing a neurological exam on himself after his coronary bypass surgery. He was so surprised by his lack of awareness of this deficit that he published a paper about it. "Even intelligent patients," he wrote, "often are surprised when a hemianopia is demonstrated, despite the fact that it has been demonstrated on numerous examinations."

The day after his surgery, Cole began to have hallucinations, in the blind half of his visual field, of people (most of whom he recognized), dogs, and horses. These apparitions did not frighten him; they "moved, danced and swirled, but their purpose was unclear." Often he hallucinated "a pony with his head cradled in my right arm"; he recognized this as his granddaughter's pony, but as with many of his hallucinations, "the colour was wrong." He always realized that these visions were unreal.

In a 1976 paper, the neurologist James Lance provided rich descriptions of thirteen hemianopic patients, and he emphasized that their hallucinations were always recognized as such, if only by their absurdity or irrelevance: giraffes and hippopotamuses sitting on one side of a pillow, visions of spacemen or Roman soldiers to one side, and so on. Other physicians have made similar reports; none of their patients ever confuses such hallucinations with reality.

I was therefore surprised and intrigued to receive the following letter from a physician in England, about his eighty-six-year-old father, Gordon H., who had long-standing glaucoma and macular degeneration. He had never had hallucinations before, but recently he had had a small stroke affecting his right occipital lobe. He was "quite sane and largely intellectually undiminished," his son wrote, but

> he has not recovered vision and retains a left hemianopia. He has, however, little awareness of his visual loss as his brain appears to fill in the missing parts. Interestingly, though, his visual hallucinations / filling in always seem to be context-sensitive or *consistent*. In other words, if he is walking in a rural setting, he can be aware of bushes and trees or distant buildings in his left visual field, which when he turns to engage his right side, he discovers are not really there. The hallucinations do, however, seem to be filled in seamlessly with his ordinary vision. If he is at his kitchen bench, he "sees" the entire bench, even to the extent of perceiving a certain bowl or plate within the left side of his vision—but which on turning disappear, because they were never really there. Yet he definitely sees a *whole* bench, with no clear separation between parts composed of hallucination and true perception.

Gordon H.'s normal visual perception to the right side, one might think, by its normalcy and detail, would immediately show up the relative poverty of the mental construct, the hallucination, on the left. But, his son asserts, he cannot tell one from the other—there is no sense of a boundary; the two halves seem continuous. Mr. H.'s case is unique, to my knowledge.[2] He has none of the outlandish, obviously out-of-context hallucinations commonly reported in hemianopia. His hallucinations blend perfectly well with his environment and seem to "complete" his missing perception.

In 1899, Gabriel Anton described a singular syndrome in which patients totally blind from cortical damage (usually from a stroke affecting the occipital lobes on both sides) seemed to be unaware of it. Such patients may be sane and intact in all other ways, but they will insist that they can see perfectly well. They will even behave as if sighted, boldly walking in unfamiliar places. If, in so doing, they collide with a piece of furniture, they will insist that the furniture has been moved, that the room is poorly lit, and so on. A patient with Anton's syndrome, if asked, will describe a stranger in the room by providing a fluent and confident, though entirely incorrect, description. No argument, no evidence, no appeal to reason or common sense is of the slightest use.

It is not clear why Anton's syndrome should produce such erroneous but unshakable beliefs. There are similar irrefutable beliefs in patients who lose the perception of their left side and the left side of space but maintain that there is nothing missing, even though we can demonstrate convincingly that they live in a hemi-universe. Such syndromes—so-called

2. In a letter to me, James Lance commented, "I have never encountered hallucination embracing information from the surroundings like Mr. H.'s."

anosognosias— occur only with damage to the right half of the brain, which seems to be especially concerned with the sense of bodily identity.

An even stranger twist was given to the matter in 1984, with the publication of a paper by Barbara E. Swartz and John C. M. Brust. Their patient was an intelligent man who had lost the sight in both eyes from retinal injuries. Normally, he recognized that he was blind and behaved as if blind. But he was also an alcoholic, and twice, while on a drinking binge, he believed that his sight had returned. Swartz and Brust wrote:

> During these episodes, he believed he could see; for example, he would walk about without asking for assistance, or he would watch television, and he claimed he could then discuss the program with friends. . . . [He] could not read the 20/800 line on a visual acuity chart, or detect a bright light or hand movements in front of his left eye. Nonetheless, he claimed that he could see, and in response to questions he offered plausible confabulations—for example describing the examining room or the appearance of the two physicians with whom he was speaking. In many particulars his descriptions were wrong, but he did not recognize that they were wrong. However, he did admit that he was also seeing things that were not really there. For example, he described the examining room as being full of little children, all wearing similar attire, some of whom were walking in and out of the room through the walls. He also described a dog in the corner eating a bone, and then noted that the walls and the floor of the room were orange. The children, dog and wall colors he recognized as hallucinations, but [he] insisted that his other visual experiences were real.

Returning to Gordon H., I would hazard a guess that damage to the right occipital lobe has produced a unilateral Anton's syndrome (though I do not know if such a syndrome has ever been described). His hallucinations (unlike those of Lance's patients) are informed and shaped by what he perceives in the intact part of his visual field, and mesh seamlessly with his intact perception to the right.

Mr. H. has only to turn his head to discover that he has been deceived, but this does not shake his conviction that he can see equally to both sides. He may, if pressed, accept the term "hallucination," but if he does so he must feel that, for him, hallucination is veridical, that he is hallucinating reality.

10

# Delirious

A s a medical student at the Middlesex Hospital in London in the 1950s, I saw many patients with delirium, states of fluctuating consciousness sometimes caused by infections with high fevers or by problems like kidney or liver failure, lung disease, or poorly controlled diabetes, all of which may produce drastic changes in blood chemistry. Some patients were delirious from medications, especially those receiving morphine or other opiates for pain. Patients with delirium were almost always on medical or surgical wards, not on neurological or psychiatric wards, for delirium generally indicates a medical problem, a consequence of something affecting the whole body, including the brain, and it disappears as soon as the medical problem has been righted.

It may be that age, even when there is full intellectual function, increases the risk of hallucinations or delirium in response to medical problems and medication—especially with the polypharmacy so often practiced in medicine today.

Working in a number of old-age homes, I sometimes see patients on a dozen or more different medications, which are liable to interact with one another in complex ways and, not uncommonly, tip the patients into delirium.[1]

We had one patient on a medical ward at the Middlesex Hospital, Gerald P., who was dying from kidney failure—his kidneys could no longer clear the toxic levels of urea building up in his blood, and he was delirious. Mr. P. had spent much of his life supervising tea plantations in Ceylon. I read this in his chart, but I could have gathered it from what he said in his delirium, for he talked nonstop, with wild associational leaps from one thought to another. My professor had said he was "talking nonsense," and at first I could make little sense of what he was saying—but the more I listened, the more I understood. I started spending as much time as I could with him, sometimes two or three hours a day. I began to see how fact and fantasy were admixed in the hieroglyphic form of his delirium, how he was reliving and at times hallucinating the events and passions of a long and varied life. It was like being privy to a dream. At first he talked to no one in particular; but

---

1. In addition to the overt delirium that may be associated with life-threatening medical problems, it is not uncommon for people to have slight delirium, so mild that it would not occur to them to consult a physician, and which they themselves may disregard or forget. Gowers, in 1907, wrote that migraine is "often attended by quiet delirium of which nothing can be subsequently recalled."

There has always been inconsistency in defining delirium, and as Dimitrios Adamis and his colleagues pointed out in their review of the subject, it has frequently been confused with dementia and other conditions. Hippocrates, they wrote, "used about sixteen words to refer to and name the clinical syndrome which we now call delirium." There was additional confusion with the medicalization of insanity in the nineteenth century, as German Berrios has noted, so that insanity was referred to as *délire chronique*. Even now the terminology is ambiguous, so that delirium is sometimes called "toxic psychosis."

once I started to ask him questions, he responded. I think he was glad that someone was listening; he became less agitated, more coherent in his delirium. He died peacefully a few days later.

In 1966, when I started practice as a young neurologist, I began working at Beth Abraham Hospital in the Bronx, a home for those with chronic diseases. One patient there, Michael F., was an intelligent man who, besides other problems, had a very damaged, cirrhotic liver, the result of a severe hepatitis infection. The little liver he had left could not cope with a normal diet, and his protein intake had to be strictly limited. Michael found this hard to take, and every so often he "cheated" by eating some cheese, which he adored. But one day, it seemed, he went too far, for he was found in a near coma. I was called at once, and when I arrived, I found Mr. F. in an extraordinary state, alternating between stupor and delirious agitation. There were brief periods when he would "come together" and show insight into what was going on. "I'm out of this world," he said at one point. "I'm stoned on protein."

When I asked him what this state felt like, he said, "like a dream, confused, sort of crazy, spaced out. But I know I'm high, as well." His attention seemed to dart about, touching on one thing and then another almost at random. He was very restless and had all sorts of involuntary movements. I had my own EEG machine at the time, and, wheeling that into Mr. F.'s room, I found that his brain waves were dramatically slowed—his EEG showed classic slow "liver waves" as well as other abnormalities. Within twenty-four hours of resuming his low-protein diet, though, Mr. F. was back to normal, as was his EEG.

Many people—especially children—experience delirium with a high fever. One woman, Erika S., recalled this in a letter to me:

> I was 11 years old and was home from school with chicken pox and a high fever. . . . During a fever spike, I experienced a frightening hallucination for what seemed like a very long time, in which my body seemed to shrink and grow. . . . With each of my breaths, my body would feel like it was swelling and swelling until I was sure that my skin would burst like a balloon. Then when it felt so excruciating, like I had suddenly grown from a normal sized child to a grotesquely fat person . . . like a person-balloon . . . I would look down at myself, sure that I would see my insides bursting out of my inadequate amount of skin, and blood pouring from enlarged orifices that could not contain my swollen body. But I would "see" my normal sized self . . . and looking would reverse the process. . . . I would feel like my body was shrinking. My arms and legs would get thinner and thinner . . . then skinny, then emaciated, then cartoon thin (like the legs on Mickey Mouse in *Steamboat Willie*) and then so pencil thin that I thought my body would disappear altogether.

Josée B. also wrote to me about her "Alice-in-Wonderland syndrome" as a child with fever. She remembered feeling "incredibly small or incredibly large and sometimes both at

the same time." She also experienced distortions in proprioception, her perception of her own body position: "One evening, I couldn't sleep in my own bed—every time I lay down
on it, I would feel I was standing tall." She had a visual hallucination, too: "Suddenly I saw cowboys who were throwing
apples at me. I jumped onto my mother's dresser to hide behind
a lipstick tube."

Another woman, Ellen R., had visual hallucinations that
took on a rhythmic, pulsing quality:

> I would "see" a smooth surface, like glass, or like the surface
> of a pond. . . . Concentric rings would spread from the center
> to the outside edges, as though a pebble had been dropped right
> in the middle. This rhythm starts slowly [but] . . . eventually
> speeds up, so that the surface is constantly agitated, and as
> this happens, my own agitation is heightened. Eventually the
> rhythm slows, the surface smooths out, and I become relieved
> and calmer myself.

Sometimes in a delirium there may be a deep humming sound
that waxes and wanes in a similar way.

While many people describe delirious swellings of body
image, Devon B., when feverish, experienced mental or intellectual swellings instead:

> What made them so strange was that they weren't sensory
> hallucinations, but a hallucination of an abstract idea . . . a
> sudden dread of a very, very large and growing number (or a
> thing, but a thing I never really defined). . . . I remember pacing
> up and down the hallway . . . in a growing state of panic and
> horror at an exponentially increasing, impossible number. . . .

My fear was that this number was violating some very basic precept of the world . . . an assumption we hold that absolutely should not be violated.

This letter made me think of the arithmetical deliria which Vladimir Nabokov went through, wrestling with impossibly large numbers, as he described in his autobiography *Speak, Memory:*

As a little boy, I showed an abnormal aptitude for mathematics, which I completely lost in my singularly talentless youth. This gift played a horrible part in tussles with quinsy or scarlet fever, when I felt enormous spheres and huge numbers swell relentlessly in my aching brain. . . . I had read . . . about a certain Hindu calculator who in exactly two seconds could find the seventeenth root of, say, 3529471145760275132301897342055-866171392 (I am not sure I have got this right; anyway the root was 212). Such were the monsters that thrived on my delirium, and the only way to prevent them from crowding me out of myself was to kill them by extracting their hearts. But they were far too strong, and I would sit up and laboriously form garbled sentences as I tried to explain things to my mother. Beneath my delirium she recognized sensations she had known herself, and her understanding would bring my expanding universe back to a Newtonian norm.

Some people feel that the hallucinations and strange thoughts of delirium may provide, or seem to provide, moments of rich emotional truth, as with some dreams or psychedelic experiences. There may also be revelations or breakthroughs of deep intellectual truth. In 1858, Alfred Russel Wallace, who had been traveling the world for a decade, collecting specimens of plants and animals and considering the problem of evolution, suddenly conceived the idea of natural selection during an attack of malarial fever. His letter to Darwin proposing this theory pushed Darwin to publish *On the Origin of Species* the following year.

Robert Hughes, in the opening of his book on Goya, writes about a prolonged delirium during his recovery from a nearly fatal car crash. He was in a coma for five weeks and hospitalized for almost seven months. In intensive care, he wrote,

> One's consciousness . . . is strangely affected by the drugs, the intubation, the fierce and continuous lights, and one's own immobility. These give rise to prolonged narrative dreams, or hallucinations, or nightmares. They are far heavier and more enclosing than ordinary sleep-dreams and have the awful character of inescapability; there is nothing outside them, and time is wholly lost in their maze. Much of the time, I dreamed about Goya. He was not the real artist, of course, but a projection of my fears. The book I meant to write on him had hit the wall; I had been blocked for years before the accident.

In this strange delirium, Hughes wrote, a transformed Goya seemed to be mocking and tormenting him, trapping him in some hellish limbo. Eventually, Hughes interpreted this "bizarre and obsessive vision":

> I had hoped to "capture" Goya in writing, and he instead imprisoned me. My ignorant enthusiasm had dragged me into a trap from which there was no evident escape. Not only could

I not do the job; my subject knew it and found my inability hysterically funny. There was only one way out of this humiliating bind, and that was to crash through. . . . Goya had assumed such importance in my subjective life that whether I could do him justice in writing or not, I couldn't give up on him. It was like overcoming writer's block by blowing up the building in whose corridor it had occurred.

Alethea Hayter, in her book *Opium and the Romantic Imagination*, writes that Piranesi, the Italian artist, was "said to have conceived the idea of his engravings of Imaginary Prisons when he was delirious with malaria," a disease he contracted

while he explored the ruined monuments of Ancient Rome . . . among the nocturnal miasmas of that marshy plain. He was bound to get malaria; and the delirious visions when they came to him may have owed something to opium as well as to a high temperature, since opium was then a normal remedy for ague or malaria. . . . The images which were born during his delirious fever were executed and elaborated over many years of fully conscious and controlled labour.

Delirium may produce musical hallucinations, as Kate E. wrote:

I was about eleven, in bed with a high fever, when I heard some heavenly music. I understood it to be a choir of angels, even though I found this odd, as I don't believe in heaven or angels and never have. So I decided it must be coming from Christmas carolers on our front doorstep below. After a minute or so, I realized it was springtime, and that I must be hallucinating.

A number of people have written to me that they have *visual* hallucinations of music, hallucinating musical notation all over the walls and ceiling. One of them, Christy C., recalled:

As a child, I ran high fevers when sick. With each spell, I would hallucinate. This was an optical hallucination involving musical notes and stanzas. I did not hear music. When the fever was high, I would see notes and clef lines, scrambled and out of order. The notes were angry and I felt unease. The lines and notes were out of control and at times in a ball. For hours, I would try to mentally smooth them out and put them in harmony or order. This same hallucination has plagued me as an adult when feverish.

Tactile hallucinations, too, can come with fever or delirium, as Johnny M. described: "When I had high fevers as a child I had very weird tactile hallucinations . . . a nurse's fingers would switch from being beautiful smooth porcelain to rough, brittle-feeling twigs or my bed sheets would go from luscious satin to drenched, heavy blankets."

Fevers are perhaps the commonest cause of delirium, but there may be a less obvious metabolic or toxic cause, as recently happened with a physician friend of mine, Isabelle R. She had had two months of increasing weakness and occasional confusion; finally she became unresponsive and was taken to the hospital, where she had a florid delirium, with hallucinations and delusions. She was convinced that a secret laboratory was hidden behind a picture on the wall of her hospital room—and that I was supervising a series of experiments on her. She was found to have extremely high levels of calcium and vitamin D (she had been taking large doses of these for her osteoporosis), and as soon as these toxic levels dropped, her delirium ceased, and she returned to normal.

Delirium is classically associated with alcohol toxicity or withdrawal. Emil Kraepelin, in his great 1904 *Lectures on Clinical Psychiatry*, included the case history of an innkeeper who developed delirium tremens from drinking six or seven liters of wine a day. He became restless and immersed in a dreamlike state in which, Kraepelin wrote,

> particular real perceptions . . . are mingled with numerous very vivid false perceptions, especially of sight and hearing. As in a dream, a whole series of the most strange and remarkable events take place with occasional sudden changes of scene. . . . Given the vivid hallucinations of sight, the restlessness, the strong tremors, and the smell of alcohol, we have all the essential features of the clinical condition called *delirium tremens.*

The innkeeper had some delusions, too, perhaps produced by his hallucinations:

> We learn, by questioning him, that he is going to be executed by electricity, and also that he will be shot. "The picture is not clearly painted," he says; "every moment someone stands now here, now there, waiting for me with a revolver. When I open my eyes they vanish." He says that a stinking fluid has been injected

into his head and both his toes, which causes the pictures [he] takes for reality. . . . He looks eagerly at the window, where he sees houses and trees vanishing and reappearing. With slight pressure on his eyes, he sees first sparks, then a hare, a picture, a washstand-set, a half-moon, and a human head, first dully and then in colours.

While deliria such as the innkeeper's may be incoherent, without any theme or connecting thread, other deliria convey the sense of a journey, or a play, or a movie, giving coherence and meaning to the hallucinations. Anne M. had such an experience after she had run a high temperature for several days. She first saw patterns whenever she closed her eyes to go to sleep; she described them as resembling Escher drawings in their sophistication and symmetry:

The initial drawings were geometric but then evolved into monsters and other rather unpleasant creatures. . . . The drawings were not in color. I was not enjoying this at all because I wanted to sleep. Once a drawing was complete it was copied so all four or six or eight quadrants of my visual field would be full of these identical pictures.

These drawings were succeeded by richly colored images that reminded her of Brueghel paintings. Increasingly, these too became full of monsters and subdivided themselves, polyopically, into a cluster of identical mini-Brueghels.

Then came a more radical change. Anne found herself in the back of "a 1950s Chinese bus on a propaganda tour of Chinese Christian churches." She recalls watching a movie on religious freedom in China projected onto the rear window of the bus.

But the viewpoint kept changing—both the movie and the bus suddenly tilted to odd angles, and it was unclear, at one point, whether a church spire she saw was "real," outside the bus, or part of the movie. Her strange journey occupied the greater part of a feverish and insomniac night.

Anne's hallucinations appeared only when she closed her eyes and would vanish as soon as she opened them.[2] But other deliria may produce hallucinations that seem to be present in the real environment, seen with the eyes open.

In 1996, I was visiting Brazil when I started to have elaborate narrative dreams with extremely brilliant colors and an almost lithographic quality, which seemed to go on all night, every night. I had gastroenteritis with some fever, and I assumed that my strange dreams were a consequence of this, compounded, perhaps, by the excitement of traveling along the Amazon. I thought these delirious dreams would come to an end when I got over the fever and returned to New York. But, if anything, they increased and became more intense than ever. They had something of the character of a Jane Austen novel, or perhaps a *Masterpiece Theatre* version of one, unfolding in a leisurely way. These visions were very detailed, with all the characters dressed, behaving, and talking as they might in *Sense and Sensibility*. (This astonished me—for I have never had much social

---

2. Just such an appearance of delirious images when closing the eyes, and their disappearance when the eyes are opened, is described by John Maynard Keynes in his memoir "Dr. Melchior":

> By the time we were back in Paris, I was feeling extremely unwell and took to my bed two days later. High fever followed. . . . I lay in my suite in the Majestic, nearly delirious, and the image of the raised pattern on the *nouveau art* wall-paper so preyed on my sensibilities in the dark that it was a relief to switch on the light and, by perceiving the reality, to be relieved for a moment from the yet more hideous pressure of its imagined outlines.

sense or sensibility, and my taste in novels inclines more to Dickens than Austen.) I would get up at intervals during the night, dab cold water on my face, empty my bladder, or make a cup of tea, but as soon as I returned to bed and closed my eyes again I was in my Jane Austen world. The dream had moved on while I was up, and when I rejoined it, it was as if the narrative had continued in my absence. A period of time had passed, events had transpired, some characters had disappeared or died, and other new ones were now on stage. These dreams, or deliria, or hallucinations, whatever they were, came every night, interfering with normal sleep, and I became increasingly exhausted from sleep deprivation. I would tell my analyst about these "dreams," which I remembered in great detail, unlike normal dreams. He said, "What's going on? You have produced more dreams in the past two weeks than in the previous twenty years. Are you *on* something?"

I said no—but then I remembered that I had been put on weekly doses of the antimalarial drug Lariam before my trip to the Amazon, and that I was supposed to take two or three further doses after my return.

I looked up the drug in the *Physician's Desk Reference*—it mentioned excessively vivid or colorful dreams, nightmares, hallucinations, and psychoses as side effects, but with an incidence of less than 1 percent. When I contacted my friend Kevin Cahill, an expert in tropical medicine, he said that he would put the incidence of excessively vivid, colorful dreams closer to 30 percent—the full-blown hallucinations or psychoses were considerably rarer. I asked him how long the dreams would go on. A month or more, he said, because Lariam has a very long half-life and would take that long to be eliminated from the body. My nineteenth-century dreams gradually faded, though they took their time doing so.

Richard Howard, the poet, was thrown into a delirium for several days following back surgery. The day after the operation, lying in his hospital bed and looking up, he saw small animals all around the edges of the ceiling. They were the size of mice but had heads like those of deer; they were vivid: solid, animal-colored, with the movements of living creatures. "I knew they were real," he said, and he was astonished when his partner, arriving at the hospital, could not see them. This did not shake Richard's conviction; he was simply puzzled as to why his partner, an artist, could be so blind (after all, he was the one who was usually so good at seeing things). The thought that he might be hallucinating did not enter Richard's mind. He found the phenomenon remarkable ("I'm not accustomed to things like a frieze of deer heads on mouse bodies"), but he accepted them as real.

The next day, Richard, who teaches literature at a university, began seeing another remarkable sight, a "pageant of literature." The physicians, nurses, and hospital staff had dressed up as literary figures from the nineteenth century, and they were rehearsing the pageant. He was very impressed by the quality of their work, although he understood that some other observers were more critical. The "actors" talked freely among themselves, and with Richard. The pageant, he could see, took place on several floors of the hospital simultaneously; the floors seemed transparent to him, so that he could watch all the levels of the performance at once. The rehearsers wanted his opinion, and he told them he thought it very attractively and intelligently done, delightful. Telling me this story six years later, he smiled, saying that even recollecting it was a delight. "It was a very privileged time," he said.

When real visitors came, the pageant would disappear, and Richard, alert and oriented, chatted with them in his usual way. But as soon as they left, the pageant recommenced. Richard is a man with an acute and critical mind, but his critical faculty, it seems, was in abeyance during his delirium, which lasted for three days, and was perhaps provoked by opiates or other drugs.

Richard is a great admirer of Henry James—and James, as it happens, also had a delirium, a terminal delirium, in December 1915, associated with pneumonia and a high fever. Fred Kaplan describes it in his biography of James:

> He had entered another imaginative world, one connected to the beginning of his life as a writer, to the Napoleonic world that had been a lifelong metaphor for the power of art, for the empire of his own creation. He began to dictate notes for a new novel, "fragments of the book he imagines himself to be writing." As if he were now writing a novel of which his own altered consciousness was the dramatic center, he dictated a vision of himself as Napoleon and his own family as the imperial Bonapartes. . . . William and Alice he grasped with his regent hand, addressing his "dear and most esteemed brother and sister." To them, to whom he had granted countries, he now gave the responsibility of supervising the detailed plans he had created for "the decoration of certain apartments, here of the Louvre and Tuileries, which you will find addressed in detail

to artists and workmen who take them in hand." . . . He was himself the "imperial eagle."

Taking down the dictation, Theodora [his secretary] felt it to be almost more than she could bear. "It is a heart-breaking thing to do, though, there is the extraordinary fact that his mind *does* retain the power to frame perfectly characteristic sentences."

This was recognized by others too—and it was said that though the master was raving, his style was "pure James" and, indeed, "late James."

Sometimes withdrawal from drugs or alcohol may cause a delirium dominated by hallucinatory voices and delusions—a delirium which is, in effect, a toxic psychosis, even though the person is not schizophrenic and has never had a psychosis before. Evelyn Waugh provided an extraordinary account of this in his autobiographical novel *The Ordeal of Gilbert Pinfold*.[3] Waugh had been a very heavy drinker for years, and at some point in the 1950s he had added a potent sleeping draft (an elixir of chloral hydrate and bromide) to the

---

3. In a prefatory note to a later edition, Waugh wrote: "Three years ago Mr. Waugh suffered a brief bout of hallucinations resembling what is here described. . . . Mr. Waugh does not deny that 'Mr. Pinfold' is largely based on himself." Thus we may accept *The Ordeal* as an autobiographic "case history" of a psychosis, an organic psychosis, albeit one written with a mastery of observation and description—and a sense of plot and suspense—that no purely medical case history has.

W. H. Auden once said that Waugh had "learned nothing" from his ordeal, but it at least enabled him to write a richly comic memoir, a new departure quite unlike anything he had written previously.

alcohol. The draft grew stronger and stronger, as Waugh wrote of his alter ego, Gilbert Pinfold: "He was not scrupulous in measuring the dose. He splashed into the glass as much as his mood suggested and if he took too little and woke in the small hours he would get out of bed and make unsteadily for the bottle and a second swig."

Feeling ill and unsteady, and with his memory occasionally playing tricks on him, Pinfold decides that a cruise to India might be restorative. His sleeping mixture runs out after two or three days, but his drinking stays at a high level. Barely has the ship got under way than he starts to have auditory hallucinations; most are of voices, but on occasion he hears music, a dog barking, the sound of a murderous beating administered by the captain of the ship and his doxy, and the sound of a huge mass of metal being thrown overboard. Visually, everything and everyone seems normal—a quiet ship with unremarkable crew and passengers, steaming quietly past Gibraltar into the Mediterranean. But complex and sometimes preposterous delusions are engendered by his auditory hallucinations: he understands, for example, that Spain has claimed sovereignty over Gibraltar and will be taking possession of the vessel, and that his persecutors possess thought-reading and thought-broadcasting machines.

Some of the voices address him directly—tauntingly, hatefully, accusingly; they often suggest that he commit suicide—although there is a sweet voice, too (the sister of one of his tormentors, he understands), who says she is in love with him, and asks if he loves her. Pinfold says he must see her, as well as hear her, but she says that this is impossible, that it is "against the Rules." Pinfold's hallucinations are exclusively auditory, and he is not "allowed" to see the speaker—for this might shatter the delusion.

Such elaborate deliria and psychoses have a top-down as well as a bottom-up quality, like dreams. They are volcano-like eruptions from the "lower" levels in the brain—the sensory association cortex, hippocampal circuits, and the limbic system—but they are also shaped by the intellectual, emotional, and imaginative powers of the individual, and by the beliefs and style of the culture in which he is embedded.

A great many medical and neurological conditions, as well as all sorts of drugs (whether taken for therapeutic purposes or for recreation), can produce such temporary, "organic" psychoses. One patient who stays most vividly in my mind was a postencephalitic man, a man of much cultivation and charm, Seymour L. (I refer to him and his hallucinations briefly in *Awakenings*). When given a very modest dose of L-dopa for his parkinsonism, Seymour became pathologically excited and, in particular, started to hear voices. One day he came up to me. I was a kind man, he said, and he had been shocked to hear me say, "Take your hat and your coat, Seymour, go up to the roof of the hospital, and jump off."

I replied that I would not dream of saying anything like that to him, and that he must be hallucinating. "Did you *see* me?" I continued.

"No," Seymour answered, "I just heard you."

"If you hear the voice again," I said to him, "look round and see if I am there. If you cannot see me, you will know it is a hallucination." Seymour pondered this briefly, then shook his head.

"It won't work," he said.

The next day he again heard my voice telling him to take his hat and his coat, go up to the roof of the hospital, and jump off, but now the voice added, "And you don't need to turn round, because I am really here." Fortunately, Mr. L. was able to resist jumping, and when we stopped his L-dopa, the voices stopped, too. (Three years later, Seymour tried L-dopa again, and this time he responded beautifully, without a hint of delirium or psychosis.)

# On the Threshold of Sleep

In 1992, I received a letter from Robert Utter, an Australian man who had heard me speak about migraine aura on
television. He wrote, "You described how some migraine
sufferers see elaborate patterns before their eyes ... and
speculated that they might be a manifestation of some deep
pattern-generating function in the brain." This reminded him
of the experience that he routinely had upon going to bed:

> This usually occurs at the moment when my head hits the
> pillow at night; my eyes close and ... I see imagery. I do not
> mean pictures; more usually they are patterns or textures, such
> as repeated shapes, or shadows of shapes, or an item from an
> image, such as grass from a landscape or wood grain, wavelets
> or raindrops ... transformed in the most extraordinary ways
> at a great speed. Shapes are replicated, multiplied, reversed in
> negative, etc. Color is added, tinted, subtracted. Textures are
> the most fascinating; grass becomes fur becomes hair follicles

becomes waving, dancing lines of light, and a hundred other variations and all the subtle gradients between them that my words are too coarse to describe.

These images and their subsequent changes appear and fade without my control. The experience is fugitive, sometimes lasting a few seconds, sometimes minutes. I cannot predict their appearance. They appear to take place not in my eye, but in some dimension of space before me. The strength of the imagery varies from barely perceptible to vivid, like a dream image. But unlike dreams, there are absolutely no emotional overtones. Though they are fascinating, I do not feel moved by them. . . . The whole experience seems to be devoid of meaning.

He wondered whether this imagery represented a sort of "idling" in the visual part of the brain, in the absence of perception.

What Mr. Utter described so vividly are not dreams but involuntary images or quasi-hallucinations appearing just before sleep—hypnagogic hallucinations, to use the term coined by the French psychologist Alfred Maury in 1848. They are estimated to occur in a majority of people, at least occasionally, although they may be so subtle as to go unnoticed.

While Maury's original observations were all of his own imagery, Francis Galton provided one of the first systematic investigations of hypnagogic hallucinations, gathering information from a number of subjects. In his 1883 book *Inquiries into Human Faculty*, he observed that very few people might at first admit to having such imagery. It was only when he sent out questionnaires stressing the common and benign qualities of these hallucinations that some of his subjects felt free to speak about them.

Galton was struck by the fact that he, too, had hypnagogic hallucinations, even though it had taken time and patience for him to realize this. "Had I been asked, before I thought of carefully trying, I should have emphatically declared that my field of view in the dark was essentially of a uniform black, subject to an occasional light-purple cloudiness and other small variations," he wrote. Once he began observing more closely, however, he saw that

> a kaleidoscopic change of patterns and forms is continually going on, but they are too fugitive and elaborate for me to draw with any approach to truth. I am astonished at their variety. . . . They disappear out of sight and memory the instant I begin to think about anything, and it is curious to me that they should often be so certainly present and yet be habitually overlooked.

Among the scores of people who responded to Galton's questionnaire was the Reverend George Henslow ("whose visions," Galton wrote, "are far more vivid than mine").[1] One of Henslow's hallucinations started with a vision of a crossbow, then of an arrow, then a flight of arrows, which changed into falling stars and then into flakes of snow. This was followed by a finely detailed vision of a rectory and then of a bed of red tulips. There were quickly changing images in which he reported visual association (for instance, arrows became stars, then snowflakes) but no narrative continuity. Henslow's imagery was extremely vivid, but it had no quality of a dream or story.

---

1. The Reverend Henslow was a son of the botanist John Stevens Henslow, who was Darwin's teacher at Cambridge and was instrumental in getting him a position aboard the *Beagle*.

Henslow emphasized how greatly these hallucinations differed from voluntary images; the latter were assembled slowly, bit by bit, like a painting, and seemed to be in the realm of everyday experience, while the former appeared spontaneously, unbidden and full-blown. His hypnagogic hallucinations were "very frequently of great beauty and highly brilliant. Cut glass (far more elaborate than I am conscious of ever having seen), highly chased gold and silver filigree ornaments; gold and silver flower-stands, etc.; elaborate colored patterns of carpets in brilliant tints."

While Galton singled out this description for its clarity and detail, Henslow was only one of many who described essentially similar visions when they were in a quiet, darkened room, ready for sleep. These visions varied in vividness, from faint imagery such as Galton himself had to virtual hallucination, though such hallucinations were never mistaken for reality.

Galton did not regard the disposition to hypnagogic visions as pathological; he thought that while a few people might experience them frequently and vividly whenever they went to sleep, most (if not all) people experienced them at least on occasion. It was a normal phenomenon, although special conditions—darkness or closing the eyes, a passive state of mind, the imminence of sleep—were needed to bring it out.

Few other scientists paid much attention to hypnagogic visions until the 1950s, when Peter McKellar and his colleagues started what was to be a decades-long investigation of near-sleep hallucinations, making detailed observations of their content and prevalence in a large population (the student body at the University of Aberdeen) and comparing them with other forms of hallucination, especially those induced by mescaline. In the 1960s, they were able to complement their phenomenological observations with EEG studies as their subjects passed from full wakefulness to a hypnagogic state.

More than half of McKellar's subjects reported hypnagogic imagery, and auditory hallucinations (of voices, bells, or animal or other noises) were just as common as visual ones. Many of my own correspondents also describe simple auditory hallucinations: dogs barking, telephones ringing, a name being called.

In his book *Upstate*, Edmund Wilson described a hypnagogic hallucination of a sort that many people share:

> I seem to hear the telephone ringing just before I am completely awake in the morning. At first, I would go to answer it, but find that it was not ringing. Now I simply lie in bed, and if the sound is not repeated, I know that it is imaginary and don't get up.

Antonella B. hears music as she is falling asleep. The first time it happened, she wrote, "I heard a really nice classical piece, played by a big orchestra, very complex and unknown." Usually, no images accompany her music, "just beautiful sounds that fill my brain up."

Susan F., a librarian, had more elaborate auditory hallucinations, as she wrote in a letter:

> For several decades, just as I am drifting off to sleep, I have heard sentences uttered. They are always grammatically correct, usually in English, and usually spoken by a man. (On a few occasions they were spoken by a woman and once in a language

I could not understand. I can recognize the differences between the Romance languages, Chinese, Korean, Japanese, Russian, and Polish, but it was none of these.) Sometimes the sentences are commands, such as "Go get me a glass of water," but at other times they are just statements or questions. During the summer of 1993, I kept a log of what I heard. Here are some of the sentences: "Once he was walking in front of me"; "This is yours, perhaps"; "Do you know what the photo looks like?"; "Mama wants some cookies"; "I smell the unicorn"; "Go get a shampoo."

What I hear bears no relationship to what I have read, seen, experienced or remembered on that day, previous day, week or year. Frequently when my husband is driving and we are on a long trip, I will nod off in the car. The sentences come very rapidly then. I will nod off for a second, hear a sentence in the twilight of waking, repeat the sentence to my husband, and then nod off again, hear another sentence in the twilight and so on, until I decide to wake up and stay awake.

In *Speak, Memory*, Nabokov provided an eloquent description of his own hypnagogic imagery, both auditory and visual:

As far back as I remember . . . I have been subject to mild hallucinations. . . . Just before falling asleep, I often become aware of a kind of one-sided conversation going on in an adjacent section of my mind, quite independently from the actual trend of my thoughts. It is a neutral, detached, anonymous voice, which I catch saying words of no importance to me whatever—an English or a Russian sentence, not even addressed to me, and so trivial that I hardly dare give samples. . . . This silly phenomenon seems to be the auditory counterpart of certain praedormitary visions, which I also know well. . . . They come and go, without the drowsy observer's participation, but are essentially different from dream pictures for he is still master of his senses. They are often grotesque. I am pestered by roguish profiles, by some coarse-featured and florid dwarf with a swelling nostril or ear. At times, however, my photisms take on a rather soothing *flou* quality, and then I see—projected, as it were, upon the inside of the eyelid—gray figures walking between beehives, or small black parrots gradually vanishing among mountain snows, or a mauve remoteness melting beyond moving masts.

Faces are especially common in hypnagogic hallucinations, as Andreas Mavromatis emphasizes in his encyclopedic book *Hypnagogia: The Unique State of Consciousness Between Wakefulness and Sleep*. He cites one man who described this in 1886; the faces, he wrote,

seem to come up out of the darkness, as a mist, and rapidly develop into sharp delineation, assuming roundness, vividness, and living reality. They fade off only to give place to others, which succeed with surprising rapidity and in enormous multitude. Formerly the faces were wonderfully ugly. They were human, but resembling animals, yet such animals as have no fellows in the creation, diabolical-looking. . . . Latterly the faces have become exquisitely beautiful. Forms and features of faultless perfection now succeed each other in infinite variety and number.

Many other descriptions stress how common it is to see faces, sometimes clusters of faces, with each face highly individuated but unrecognizable. F. E. Leaning, in her 1925 paper on hypnagogia, speculated that such an emphasis on faces "almost suggests that there is a special 'face-seeing' propensity in the mind." Leaning's "propensity," we now know, has its anatomical substrate in a specialized portion of the visual cortex, the fusiform face area. Dominic ffytche and his colleagues have shown in fMRI studies that it is precisely this area in the right hemisphere which is activated when faces are hallucinated.

Activation of a homologous area in the left hemisphere may produce lexical hallucinations—of letters, numbers, musical notation, sometimes words or pseudowords, or even sentences. One of Mavromatis's subjects put it this way: "When dozing or before going to sleep . . . I appear to be reading a book. I see the print clearly and distinguish the words, but the words rarely seem to have any particular significance. The books I appear to be reading are never books with which I am familiar, but frequently deal with whatever subject I have been reading during the day."

(While hypnagogic images of faces and places are usually unrecognizable, there is a distinct category of hypnagogia which McKellar and Simpson call "perseverative": hallucinations or recurrent images of something one has been exposed to earlier in the day. If, for example, one has been driving all day, one may "see" a hedgerow or line of trees continually unfurling before one's closed eyes.)

Hypnagogic imagery may be faint or colorless, but it often has brilliant and highly saturated color. Ardis and McKellar, in a 1956 paper, cited a case in which the subject described "colors of the spectrum intensified as though bathed in the fiercest sunlight." They compared this, as others have, to the exaggeration of color with mescaline. In hypnagogic hallucinations, luminosity or outlines may also seem to be abnormally distinct, with shadows or furrows exaggerated—sometimes such exaggerations go with cartoonlike figures or scenes. Many people speak of an "impossible" clarity or a "microscopic" detail in their hypnagogic visions. Images may seem finer-grained than perception itself, as if the inner eye has an acuity of 20/5 rather than 20/20 (this hyperacuity is a feature common to many types of visual hallucination).

One may "see" a constellation of images in hypnagogia—a landscape in the middle, a face erupting in the upper left corner, a complex geometric pattern around the edge—all present simultaneously and all evolving or metamorphosing in their own ways, a sort of multifocal hallucination. Many people describe hallucinatory polyopia, multiplications of objects or figures (one of McKellar's subjects saw a pink cockatoo, then hundreds of pink cockatoos talking to each other).

Figures or objects may suddenly zoom towards one, getting larger and more detailed, then retreat. Hypnagogic images, often compared to snapshots or slides, flash into consciousness, hold for a second or two, then disappear; they may be replaced by other images that seem to have no connection or apparent association to one another.

Hypnagogic visions may seem like something from "another world"—this phrase is used again and again by people describing their visions. Edgar Allan Poe stressed the fact that his own hypnagogic images were not only unfamiliar but unlike anything he had ever seen before; they had "the absoluteness of novelty."[2]

Most hypnagogic images are not like true hallucinations: they are not felt as real, and they are not projected into external space. And yet they have many of the special features of hallucinations—they are involuntary, uncontrollable, autonomous; they may have preternatural colors and detail and undergo rapid and bizarre transformations unlike those of normal mental imagery.

There is something about the rapid and spontaneous transformations specific to hypnagogic imagery that suggests the brain is "idling," as my correspondent Mr. Utter suggested. Neuroscientists now tend to speak of "default networks" in the brain, which generate their own images. Perhaps one may also venture the term "play" and think of the visual cortex playing with every permutation, playing with no goal, no focus, no meanings—a random activity or perhaps an activity with so many microdeterminants that no pattern is ever repeated. Few phenomena give such a sense of the brain's creativity and computational power as the almost infinitely var-

---

2. Feeling that hypnagogic hallucinations could extend and enrich the imagination, Poe would jerk himself suddenly to full wakefulness while hallucinating, so that he could make note of the extraordinary things he saw, and he often brought these into his poems and short stories. Poe's great translator, Baudelaire, was also fascinated by the unique quality of such visions, especially if they were potentiated by opium or hashish. A whole generation in the early nineteenth century (including Coleridge and Wordsworth, as well as Southey and De Quincey) was influenced by such hallucinations. This is explored by Alethea Hayter in her book *Opium and the Romantic Imagination* and by Eva Brann in her magisterial *The World of the Imagination: Sum and Substance*.

ied, ever-changing torrent of patterns and forms which may be seen in hypnagogic states.

Although Mavromatis writes of hypnagogia as "the unique state of consciousness between wakefulness and sleep," he sees affinities with other states of consciousness—those of dreams, meditation, trance, and creativity—as well as the altered modes of consciousness in schizophrenia, hysteria, and some drug-induced states. Although hypnagogic hallucinations are sensory (and thus cortical, being produced by the visual cortex, auditory cortex, etc.), he feels that the initiating processes may be in the more primitive, subcortical parts of the brain, and this, too, is something that hypnagogia may share with dreams.

And yet the two are quite distinct. Dreams come in episodes, not flashes; they have a continuity, a coherence, a narrative, a theme. One is a participant or a participant-observer in one's dreams, whereas with hypnagogia, one is merely a spectator. Dreams call on one's wishes and fears, and they often replay experiences from the previous day or two, assisting in the consolidation of memory. They sometimes seem to suggest the solution to a problem; they have a strongly personal quality and are determined mostly from above—they are largely "top-down" creations (although, as Allan Hobson argues, with a wealth of supporting evidence, they also employ "bottom-up" processes). In contrast, hypnagogic imagery or hallucination, with its largely sensory qualities—enhanced or exaggerated color and detail and outlines, luminosity, distortions, multiplications, and zoomings—and its detachment from personal experience, is overwhelmingly a "bottom-up" process. (But this is a simplification, for given the two-way traffic at every level in the nervous system, most processes are both top-down

and bottom-up.) Hypnagogia and dreaming are both extraordinary states of consciousness, as different from each other as
they are different from waking consciousness.

Hypnopompic hallucinations—those that may come upon
waking—are often profoundly different in character
from hypnagogic hallucinations.[3] Hypnagogic hallucinations,
seen with closed eyes or in darkness, proceed quietly and fleet-
ingly in their own imaginative space and are not usually felt
to be physically present in one's room. Hypnopompic hallucinations are often seen with open eyes, in bright illumination;
they are frequently projected into external space and seem to
be totally solid and real. They sometimes give amusement or
pleasure, but they often cause distress or even terror, for they
may seem charged with intentionality and ready to attack
the just-wakened hallucinator. There is no such intentionality with hypnagogic hallucinations, which are experienced as
spectacles unrelated to the hallucinator.

While hypnopompic hallucinations are only occasional
with most people, they may occur frequently in some, as is
the case with Donald Fish, an Australian man whom I met in
Sydney after he wrote to me about his vivid hallucinations:

> I wake from a calm sleep and perhaps a fairly normal dream with
> a shock, and there before me is a creature that even Hollywood
> couldn't create. The hallucinations fade in about ten seconds,
> and I can move when I have them. In fact I usually jump about

---

3. Hypnopompic hallucinations are far less common than hypnagogic
ones, and some people have hypnagogic hallucinations upon awakening,
or hypnopompic ones while falling asleep.

a foot into the air and scream.... The hallucinations are becoming worse—now about four a night—I am now becoming terrified of going to bed. The following are some examples of what I see:

A huge figure of an angel standing over me next to a figure
     of death in black.
A rotting corpse lying next to me.
A huge crocodile at my throat.
A dead baby on the floor covered in blood.
Hideous faces laughing at me.
Giant spiders—very frequent.
Huge hand over my face. Also one on the floor five feet
     across.
Drifting spider webs.
Birds and insects flying into my face.
Two faces looking at me from under a rock.
Image of myself—only looking older—standing by the bed
     in a suit.
Two rats eating a potato.
A mass of different colored flags descending onto me.
Ugly-looking primitive man lying on floor covered in tufts
     of orange hair.
Shards of glass falling on me.
Two wire lobster pots.
Dots of red, increasing to thousands like spattered blood.
Masses of logs falling on me.

It is often said that hypnagogic and hypnopompic hallucinations are more vivid and most easily remembered in childhood, but Mr. Fish's hallucinations have been lifelong—they started when he was eight, and he is now over eighty. Why he should be so prone to hypnopompic hallucinations is a mystery. Although he has had thousands of hypnopompic hallucinations, he has been able to live a full life and function consistently at a high creative level. A graphic designer and visual artist with a brilliant imagination, he sometimes finds inspiration in his surreal hallucinations.

While Mr. Fish's hypnopompic imagery is extreme in its frequency (and very distressing to him), it is not atypical in character. Elyn S. wrote to me about her own hypnopompic images:

> The most typical one would involve me sitting up in bed and seeing a person—often an old lady—staring at me at some distance from the foot of my bed. (I imagine that such hallucinations are thought to be ghosts by some people—but not by me.) Other examples are seeing a foot-wide spider crawling up my wall; seeing fireworks; and seeing a little devil at the foot of my bed riding a bicycle in place.

A powerfully persuasive form of hallucination, not explicitly sensory at all, is the feeling of the "presence" of someone or something nearby, a presence that may be felt as malevolent or benign. The sense of conviction that someone is there can be irresistible at such times.

For me, hypnopompic experiences are usually more auditory than visual, and they take a variety of forms. Sometimes they are persistences of dreams or nightmares. On one occasion I heard a scratching sound in the corner of the room. I paid little attention at first, thinking it was just a mouse in the walls. But the scratching grew louder and louder and began to frighten me. Alarmed, I flung a pillow into the corner. But the action (or, rather, the imagined action) of flinging fully awakened me, and I opened my eyes to find that I was in my own bedroom, not the hospital-like room of my dream. But the scratching sound continued, loud and utterly "real," for several seconds after I woke.

I have had musical hallucinations (when taking chloral hydrate as a sleeping aid) which were continuations of dream music into the waking state—once with a Mozart quintet. My normal musical memory and imagery is not that strong—I am quite incapable of hearing every instrument in a quintet, let alone an orchestra—so the experience of hearing the Mozart, hearing every instrument, was a startling (and beautiful) one. Under more normal conditions I experience a hypnopompic state of heightened (and somewhat uncritical) musical sensibility; whatever music I hear in this state delights me. This happens almost every morning when I am awoken by my clock radio, which is tuned to a classical station. (An artist friend describes a similar enhancement of color and texture when he lies in bed after first opening his eyes in the morning.)

Recently, I had a startling and rather moving visual hallucination. I cannot recollect what I was dreaming, if indeed I was dreaming, but when I awoke I saw my own face—or, rather, my face as it was when I was forty, black-bearded, smiling rather shyly. The face was about two feet away, life-sized, in faint, unsaturated pastel color, poised in midair; it seemed to look at me with curiosity and affection, and then, after about five seconds, it faded out. It gave me an odd, nostalgic sense of continuity with my younger self. As I lay in bed, I wondered whether, when young, I had ever had a vision of my present, almost eighty-year-old face, a hypnopompic "hello" across forty years.

While we may have the most fantastical and surreal experiences in our dreams, we accept these because we are enveloped in our dream consciousness, and there is no critical consciousness outside it (the rare phenomenon of lucid dreaming is an exception). When we awake we can remember only fragments, a tiny fraction of our dreams, and can easily dismiss these as "just a dream."

Hallucinations, in contrast, are startling and apt to be remembered in great detail—this is one of the central contrasts between sleep-related hallucinations and dreaming. My colleague Dr. D. has had only one hypnopompic hallucination in his life, and it occurred thirty years ago. But he retains the most vivid memory of it:

> It was a relaxed summer night. I awoke around 2 AM, as I sometimes do in the middle of the night, and next to me, standing six and a half feet tall, was an imposing Native American. A huge man, muscles chiseled, black hair and black eyes. I realized, seemingly simultaneously, that if he wanted to kill me there was nothing I could do, and that he must not be real. Yet he was standing there, like a statue yet very alive. My mind flashed about—how could he have gotten into the house? . . . Why was he motionless? . . . This can't be real. Yet his presence frightened me. He became diaphanous after five or ten seconds, gently vaporizing into invisibility.[4]

4. Spinoza, in the 1660s, described a similar hallucination in a letter to his friend Peter Balling:

> When one morning, after the day had dawned, I woke up from a very unpleasant dream, the images, which had presented them-

Given the outlandish quality of some hypnopompic images, their often terrifying emotional resonance, and perhaps the heightened suggestibility that may go with such states, it is very understandable that hypnopompic visions of angels and devils may engender not only wonder or horror but belief in their physical reality. Indeed, one must wonder to what degree the very idea of monsters, ghostly spirits, or phantoms originated with such hallucinations. One can easily imagine that, coupled with a personal or cultural disposition to believe in a disembodied, spiritual realm, these hallucinations, though they have a real physiological basis, might reinforce a belief in the supernatural.

The term "hypnopompic" was introduced in 1901 by F. W. H. Myers, an English poet and classicist who was fascinated by the emerging study of psychology. He was a friend of William James's and a founding member of the Society for Psychical Research, where he sought to connect the abnormal and paranormal with normal psychological function. Myers's work was highly influential.

Living in the late nineteenth century, a time in which séances and mediums were all the rage, Myers wrote extensively of ghosts, apparitions, and phantoms. Like many of his contemporaries, he believed in the idea of life after death, but he tried to place it in a scientific context. Although he felt that

---

selves to me in sleep, remained before my eyes just as vividly as though the things had been real, especially the image of a certain black and leprous Brazilian whom I had never seen before. This image disappeared for the most part when, in order to divert my thoughts, I cast my eyes on a book or something else. But as soon as I lifted my eyes again, without fixing my attention on any particular object, the same image of this same negro appeared with the same vividness again and again, until the head of it finally vanished.

experiences likely to be interpreted as supernatural visitations were especially apt to occur in hypnopompic states, he also believed in the objective reality of a spiritual or supernatural realm, to which the mind might be given brief access in various physiological states, such as dreaming, hypnopompic states, trance states, and certain forms of epilepsy. But at the same time, he thought that hypnopompic hallucinations might be fragments of dreams or nightmares persisting into the waking state—in effect, waking dreams.

Yet reading Myers's 1903 two-volume *Human Personality and Its Survival After Bodily Death*, as well as *Phantasms of the Living*, the compilation of case histories he and his colleagues (Gurney et al.) published in 1886, one feels that the majority of "psychical" or "paranormal" experiences described are, in fact, hallucinations—hallucinations arising in states of bereavement, social isolation, or sensory deprivation, and above all in drowsy or trancelike states.

My colleague Dr. B., a psychotherapist, related the following story to me, about a ten-year-old boy who woke one morning "to find a woman dressed in blue hovering at the foot of his bed, surrounded by radiant light":

> She introduced herself as his "guardian angel," speaking in a soft, gentle voice. The child was terrified, and turned on the light beside his bed, expecting the image to disappear. The woman remained suspended in the air, however, and he ran from the room, awakening his parents.
>
> His parents framed the experience as a dream, trying to reassure the child. He was unconvinced, unable to make sense of the event. His family had no religious background, and he found the image of the angel alien. He began to experience

a pervasive sense of dread and developed insomnia, fearful that he would awaken to find the woman again. His parents and teachers described him as agitated and distracted, and he increasingly withdrew from relationships with peers and activities. His parents called their pediatrician, who referred the child for psychiatric evaluations and psychotherapy.

The child had no prior history of problems in functioning, sleep disorder, or physical illness, and he appeared to be well-adjusted. He made effective use of therapeutic consultations, where he continued to . . . make sense of what had happened, coming to understand the event as a type of hallucination that commonly occurs following arousal from sleep.

Dr. B. added, "Although there would appear to be a high prevalence of hypnopompic hallucinations among healthy, well-adjusted persons, they are potentially traumatic, and it is crucial to explore the *meaning* and implications of such phenomena for the individual."

Experiences so far out of the ordinary constitute a severe challenge to one's world picture, one's belief system—how can they be explained? What do they mean? One sees poignantly with this young patient how reason itself can be rocked by such nighttime visions, which insist on their own reality.

## 12

# Narcolepsy and Night Hags

Sometime in the late 1870s, Jean-Baptiste-Édouard Gélineau, a French neurologist from a wine-making family, had occasion to examine a thirty-eight-year-old wine merchant who had been having attacks of sudden, brief, irresistible sleep for two years. By the time he came to Gélineau, he was having as many as two hundred a day. He sometimes fell asleep in the middle of a meal, the knife and fork slipping from his fingers; he might drop off in the middle of a sentence or as soon as he had been seated in a theater. Intense emotions, sad or happy, often precipitated his sleep attacks and also episodes of "astasia," in which there was a sudden loss of muscular strength and tone, so that he would fall helplessly to the ground, while remaining perfectly conscious. Gélineau regarded this conjunction of narcolepsy (a term he coined) and astasia (we now call it cataplexy) as a new syndrome—one with a neurological origin.[1]

---

1. Bill Hayes, in his book *Sleep Demons*, cites an even earlier reference to irresistible, overwhelming sleepiness and probable cataplexy—"It falls

In 1928 a New York physician, Samuel Brock, presented a broader view of narcolepsy, describing a young man of twenty-two who was prone not only to sudden sleep attacks and cataplexy but also a paralysis, with the inability to talk or move, following his sleep attacks. In this state of sleep paralysis (as the condition was later to be named), he had vivid hallucinations, which he experienced at no other time. Though Brock's case was described in a contemporary (1929) review of narcolepsy as "unique," it soon became apparent that sleep paralysis and the hallucinations associated with it were far from uncommon and should be regarded as integral features of a narcoleptic syndrome.

It is now known that the hypothalamus secretes "wakefulness" hormones, orexins, and that these are deficient in people who have congenital narcolepsy. Damage to the hypothalamus, from a head injury or a tumor or disease, can also cause narcolepsy later in life.

Full-blown narcolepsy can be incapacitating if untreated, but it is mercifully rare, affecting perhaps one person in two thousand. (Milder forms may be appreciably commoner.) People with narcolepsy are apt to feel embarrassed, isolated, or misunderstood (as with Gélineau's patient, who was regarded as a drunk), but awareness is spreading, in part because of organizations such as the Narcolepsy Network.

Despite this, narcolepsy often goes undiagnosed. Jeanette B. wrote to me that her narcolepsy had not been diagnosed until she was an adult. In elementary school, she said, "I thought I

upon them in the midst of mirth"—from a little-known 1834 book, *The Philosophy of Sleep,* by the Scottish physician Robert Macnish.

had schizophrenia, because of my hypnagogic hallucinations. I even wrote a paper on schizophrenia in sixth grade (never mentioning that I thought that was my problem)." Much later, when she went to a narcolepsy support group, she wrote, "I was astounded to find that many in the group not only had hallucinations, but the very same hallucinations as I did!"

When I heard recently that the New York chapter of the Narcolepsy Network was due to have a meeting, I asked if I might come along to listen to members discuss their experiences and to talk with some of them myself. Cataplexy— the sudden, complete loss of muscle tone with emotion or laughter—affected many at this meeting, and it was freely discussed. (Cataplexy, indeed, can scarcely be hidden. I spoke to one man, by chance a friend of the comedian Robin Williams's, who said that whenever he met Robin, he would lie down on the ground preemptively; otherwise, he was sure to fall down in a fit of laughter-induced cataplexy.) But hallucinations were another matter: people often hesitate to admit to them, and there was little open discussion of the subject, even in a room full of narcoleptics. Nonetheless, many people later wrote to me about their hallucinations, including Sharon S., who described her own experience:

I wake on my stomach to the sensation that the mattress is breathing. I cannot move and the terror sets in as I "see" the marbled grey skin with sparse black hairs underneath me. I am sprawled on the back of a walking elephant. . . . The absurdity of my hallucinations causes me to collapse with cataplexy. . . . [Another time] as I am waking from a nap I "see" myself in the corner of the bedroom. . . . I am close to the ceiling, slowly floating to the floor by parachute. During the hallucination it seemed perfectly normal and I am left with a very peaceful, serene feeling.

Sharon has also had hallucinations while driving:

[I am driving] to work, and getting increasingly sleepy; suddenly, the road ahead rises up in front of me and hits me in the face. It is so realistic. I jerk my head back. It certainly woke me up. This experience is different from my other hallucinations in that my eyes were open and I was seeing my actual surroundings, but with distortion.

While most of us have a robust sleep-wake cycle, with sleep occurring predominantly at night, people with narcolepsy can have dozens of "microsleeps" (some lasting for only a few seconds) and "in-between states" each day—and any or all of these may be charged with intensely vivid dreams, hallucinations, or some almost-indistinguishable fusion of the two. Sudden, narcolepsy-like sleep without cataplexy may also occur in toxic states or with various medications (especially sedatives), and there is often some tendency to it with aging, in the dozing or nodding off of the elderly into brief, dream-charged sleeps.

I have these increasingly often myself. Once, while reading Gibbon's autobiography in bed—this was in 1988, when I was thinking and reading a great deal about deaf people and their use of sign language—I found an amazing description by Gibbon of seeing a group of deaf people in London in 1770, immersed in an animated sign discourse. I immediately thought that this would make a wonderful footnote for the book I was writing, but when I came to reread Gibbon's description, it was not there. I had hallucinated or perhaps dreamt it, in a flash, between two sentences of text.

Stephanie W. had her first narcoleptic hallucination when she was five, walking home from kindergarten. She wrote to me that her hallucinations frequently occur during the daytime, and she presumes they happen before or after very short microsleeps:

> However . . . I am not able to detect that a microsleep has occurred unless something in my environment noticeably "jumps" forward or changes in some way—as it did, for example, when I still drove a car and would find that my vehicle had unaccountably leapt forward on the road during a microsleep. . . . Prior to treatment for narcolepsy, I had many periods during which I experienced hallucinations on a daily basis. . . . Some were utterly benign: an "angel" which would appear periodically over a particular highway exit . . . hearing a person whispering my name repeatedly, hearing a knock at the door which no one else hears, seeing and feeling ants walking on my legs. . . . Some were terrifying [like the] experience of visually seeing the people before me take on the appearance of being dead. . . .
>
> It was especially difficult as a child to be experiencing things that the people around me did not also sense. The attempts that I remember making to talk with adults or other kids about what was going on repeatedly elicited anger and suspicion that I was "crazy" or lying. . . . It got easier as an adult. (Although when I was treated within the mental health system, I was told that I had "Psychosis with unusually strong reality testing.")

Receiving the correct diagnosis—narcolepsy—was deeply reassuring to Stephanie W., as was meeting others with similar hallucinations in the Narcolepsy Network.[2] With this diagnosis and the prescription of effective medication, she feels there has been a complete change in her life.

Lynn O. wished that her doctors had told her earlier that her hallucinations were part of a narcoleptic syndrome. Prior to her diagnosis, she wrote,

> These episodes happened frequently enough throughout my life that instead of suspecting a sleep disorder, I suspected paranormal activity in my life. Are there many people who integrate the experiences in this manner? Had I been better educated about this disorder, perhaps instead of suspecting I was being interfered with, haunted, spiritually challenged or perhaps mentally ill, I would have sought more constructive help earlier in life. I am now forty-three years old. And I have found a new peace in life in realizing many of these experiences have had to do with this disorder.

In a later letter, she observed, "I find myself in the fresh stage of having to reevaluate many of my 'paranormal' experiences, and I find I am having to reintegrate a new view of the world based on my new diagnosis. It is like letting go of childhood or, rather, letting go of a mystical, almost magical view of the world. I must say, perhaps I am experiencing a touch of mourning."

---

2. A key figure in the narcolepsy world is Michael Thorpy, a physician whose many books on narcolepsy and other sleep disorders have grown out of a lifetime of experience directing a sleep disorders clinic at Montefiore Medical Center in the Bronx.

Many people with narcolepsy have auditory or tactile hallucinations along with visual ones, as well as complex bodily feelings. Christina K. is prone to sleep paralysis, and often her hallucinations go with this, as in the following episode:

> I had just lain down in bed, and after a few rounds of changing positions I ended up face down. Almost immediately I felt my body go more and more numb. I tried to "pull" myself out of it, but I was already too deep into the paralysis. Then it was almost as if someone sat down on my back, pressing me deeper into the mattress . . . the weight on my back got heavier and heavier, and I was still not able to move. [Then] the thing on my back got off and laid down next to me. . . . I could feel it lying beside me, breathing. I got so scared and thought that this couldn't be anything other than real . . . because I had been awake all along. It felt like an eternity before I managed to turn my head towards it. Then I laid eyes on an abnormally tall man in a black suit. He was greenishly pale, sick-looking, with a shock-ridden look in the eyes. I tried to scream, but was unable to move my lips or make any sounds at all. He kept staring at me with his eyes almost popping out when all of a sudden he started shouting out random numbers, like FIVE-ELEVEN-EIGHT-ONE-THREE-TWO-FOUR-ONE-NINE-TWENTY, then laughed hysterically. . . . I started feeling able to move again, and as I came back to a normal state the image of the man became more and more blurry until he was gone and I was able to get up.

Another correspondent, J.D., also described the hallucinations associated with sleep paralysis, including the feeling of pressure on her chest:

Sometimes I would see things like huge centipedes or caterpillars crawling all over my ceiling. Once I thought my cat was on the shelf in my room. She seemed to be rolling around and turning into a rat. The worst was when I would hallucinate that a spider was on my chest. I couldn't move. I would try to scream. I am TERRIFIED of spiders.

On one occasion, she had a hallucination resembling an out-of-body experience:

I hallucinated that my body floated up to the ceiling towards the end of my bed, and then all of a sudden my body quickly dropped through the floor to the first level of the house and then dropped through that floor and into the basement. I could see everything in each room. The floors did not seem to break when I went through them. I just passed through them.

There was little physiological understanding of sleeping, dreaming, or sleep disorders until 1953, when Eugene Aserinsky and Nathaniel Kleitman at the University of Chicago discovered REM sleep—a distinctive stage of sleep with characteristic rapid eye movements, as well as characteristic EEG changes. They also found that if their subjects were woken during REM sleep, they would always report that they had been dreaming. It seemed, then, that dreaming was correlated with REM sleep.[3] In REM sleep the body is paralyzed, except for shallow breathing and eye movements. Most people

---

3. This simple equation had to be modified later, when it was found that dreams—albeit of a somewhat different kind—could also occur in non-REM sleep.

enter the REM stage ninety minutes or so after falling asleep, but people with narcolepsy (or those with sleep deprivation) may fall into REM at the very onset of sleep, plunging suddenly into dreaming and sleep paralysis; they may also wake at the "wrong" time, so that the dreamlike visions and the loss of muscle control characteristic of REM sleep persist into the waking state. Even though the person is wide awake, he may be assaulted by dream- or nightmare-like hallucinations, made even more terrifying by an inability to move or speak.

But one does not have to have narcolepsy to experience sleep paralysis with hallucinations—indeed, J. A. Cheyne and his colleagues at the University of Waterloo have shown that somewhere between a third and half of the general population has had at least occasional episodes of this, and even a single episode may be unforgettable.

Cheyne et al. explored and categorized a huge range of sleep-paralysis-related phenomena, based on reports from three hundred student subjects as well as a large and varied population who responded to an internet questionnaire. They concluded that isolated sleep paralysis (that is, sleep paralysis without narcolepsy), being relatively common, "constitutes a unique natural laboratory for the study of hallucinoid experiences" but stressed that such hallucinations cannot be compared to ordinary hypnagogic or hypnopompic experiences. The hallucinations accompanying isolated sleep paralysis, they wrote, are "substantially more vivid, elaborate, multimodal and terrifying," and therefore more likely to have a radical impact on anyone who experiences them. These hallucinations may be visceral, auditory, or tactile as well as visual and are accompanied by a feeling of suffocation or pressure on the chest, the sense of a malignant presence, and an overall sense of absolute helplessness and abject terror. These, of course, are the cardinal qualities of the nightmare, in its original sense.

The "mare" in "nightmare" originally referred to a demonic woman who suffocated sleepers by lying on their chests (she was called "Old Hag" in Newfoundland). Ernest Jones, in his monograph *On the Nightmare*, emphasized that nightmares were radically different from ordinary dreams in their invariable sense of a fearful presence (sometimes astride the chest), difficulty breathing, and the realization that one is totally paralyzed. The term "nightmare" is often used now to describe any bad dream or anxiety dream, but the real night-mare has dread of a wholly different order; Cheyne speaks of "the ominous numinous" here. He suggests that the term for the night-mare proper be spelled with a hyphen, and this convention has been adopted by other workers in the field.

Shelley Adler, in her book *Sleep Paralysis: Night-mares, Nocebos, and the Mind-Body Connection*, also brings out the extreme nature of the sense of terror and doom that makes the experience of sleep paralysis unlike any other. She emphasizes that night-mares, unlike dreams, occur when one is awake—but awake in a partial or dissociated way; in this sense, the term "sleep" paralysis is misleading. The terror of this state is heightened by the shallow breathing of REM sleep and a rapid or irregular heartbeat, which can go with extreme excitement. Such overpowering fear and its physiological accompaniments can even be fatal, especially if there is a cultural tradition that associates sleep paralysis with death. Adler studied a group of Hmong refugees from Laos who had immigrated to central California in the late 1970s and were not always able to perform their traditional religious rites during the upheaval of genocide and relocation. In Hmong culture, there is a strong belief that night-mares can be fatal; this evil expectation, or nocebo, apparently contributed to the sudden unexplained nocturnal deaths of almost two hundred Hmong immigrants (mostly young and in good health) in the late 1970s and early 1980s. Once they were more assimilated and the old beliefs lost their power, the sudden deaths stopped.

The folklore of every culture includes supernatural figures like the incubus and succubus, which assault the sleeper sexually, or the Old Hag, which paralyzes its victims and sucks their breath away. Such images seem to be universal—indeed, there is a remarkable similarity of such figures in widely disparate cultures, although there are local variations of every sort. Hallucinatory experiences, whatever their cause, generate a world of imaginary beings and abodes—heaven, hell, fairyland. Such myths and beliefs are designed to clarify and reassure and, at the same time, to frighten and warn. We make narratives for a nocturnal experience which is common, real, and physiologically based.

When traditional figures—devils, witches, or hags—are no longer believed in, new ones—aliens, visitations from "a previous life"—take their place. Hallucinations, beyond any other waking experience, can excite, bewilder, terrify, or inspire, leading to the folklore and the myths (sublime, horrible, creative, and playful) which perhaps no individual and no culture can wholly dispense with.

# 13

# The Haunted Mind

In Charles Bonnet syndrome, sensory deprivation, parkinsonism, migraine, epilepsy, drug intoxication, and hypnagogia, there seems to be a mechanism in the brain that generates or facilitates hallucination—a primary physiological mechanism, related to local irritation, "release," neurotransmitter disturbance, or whatever—with little reference to the individual's life circumstances, character, emotions, beliefs, or state of mind. While people with such hallucinations may (or may not) enjoy them as a sensory experience, they almost uniformly emphasize their meaninglessness, their irrelevance to events and issues of their lives.

It is quite otherwise with the hallucinations we must now consider, which are, essentially, compulsive returns to a past experience. But here, unlike the sometimes moving but essentially trivial flashbacks of temporal lobe seizures, it is the significant past—beloved or terrible—that comes back to haunt the mind—life experiences so charged with emotion that they make an indelible impression on the brain and compel it to repetition.

The emotions here can be of various kinds: grief or longing for a loved person or place from which death or exile or the passage of time has separated one; terror, horror, anguish, or dread following deeply traumatic, ego-threatening or life-threatening events. Such hallucinations may also be provoked by overwhelming guilt for a crime or sin that, perhaps belatedly, the conscience cannot tolerate. Hallucinations of ghosts—revenant spirits of the dead—are especially associated with violent death and guilt.

Stories of such hauntings and hallucinations have a substantial place in the myths and literature of every culture. Thus Hamlet's murdered father appears to him ("In my mind's eye, Horatio") to tell him how he was murdered and must be avenged. And when Macbeth is plotting the murder of King Duncan, he sees a dagger in midair, a symbol of his intention and an incitement to action. Later, after he has had Banquo killed for threatening to expose him, he has hallucinations of Banquo's ghost; while Lady Macbeth, who has smeared Duncan's blood over his slain grooms, "sees" the king's blood and smells it, ineradicable, on her hands.[1]

---

1. Many of H. G. Wells's short stories also involve guilt hallucinations. In "The Moth," a zoologist who feels himself responsible for the death of his lifelong rival is haunted and finally driven mad by a giant moth that no one else can see, a moth of a genus unknown to science; but in his lucid moments, he jokes that it is the ghost of his deceased rival.

Dickens, a haunted man himself, wrote five books on this theme, the best known of these being *A Christmas Carol*. And in *Great Expectations*, he provides a dramatic account of Pip's vision after his first, horrified encounter with Miss Havisham:

Any consuming passion or threat may lead to hallucinations in which an idea and an intense emotion are embedded. Especially common are hallucinations engendered by loss and grief—particularly following the death of a spouse after decades of togetherness and marriage. Losing a parent, a spouse, or a child is losing a part of oneself; and bereavement causes a sudden hole in one's life, a hole which—somehow—must be filled. This presents a cognitive problem and a perceptual one as well as an emotional one, and a painful longing for reality to be otherwise.

I never experienced hallucinations after the deaths of my parents or my three brothers, though I often dreamt of them. But the first and most painful of these losses was the sudden death of my mother in 1972, and this led to persistent illusions over a period of months, when I would mistake other people in the street for her. There was always, I think, some similarity of appearance and carriage behind these illusions, and part of me, I suspect, was hyper-alert, unconsciously searching for my lost parent.

Sometimes bereavement hallucinations take the form of a

---

I thought it a strange thing then, and I thought it a stranger thing long afterwards. I turned my eyes—a little dimmed by looking up at the frosty light—towards a great wooden beam in a low nook of the building near me on my right hand, and I saw a figure hanging there by the neck. A figure all in yellow white, with but one shoe to the feet; and it hung so that I could see that the faded trimmings of the dress were like earthy paper, and that the face was Miss Havisham's, with a movement going over the whole countenance as if she were trying to call me. In the terror of seeing the figure, and in the terror of being certain that it had not been there a moment before, I at first ran from it, and then ran towards it. And my terror was greatest of all when I found no figure there.

voice. Marion C., a psychoanalyst, wrote to me about "hearing" the voice (and, on a subsequent occasion, the laugh) of her dead husband:

> One evening I came home from work as always to our big empty house. Usually at that hour Paul would have been at his electronic chessboard playing over the game in the *New York Times.* His table was out of sight of the foyer, but he greeted me in his familiar way: "Hello! You're back! Hi!" . . . His voice was clear and strong and true; just the way it was when he was well. I "heard" it. It was as if he were actually at his chess table and actually greeting me once more. The other part was that, as I said, I couldn't see him from the foyer, yet I did. I "saw" him, I "saw" the expression on his face, I "saw" how he moved the pieces, I "saw" him greet me. That part was like one sees in a dream: as if I were seeing a picture or a movie of an event. But the speech was live and real.

Silas Weir Mitchell, working with soldiers who had lost limbs in the Civil War, was the first to understand the neurological nature of phantom limbs—they had previously been regarded, if at all, as a sort of bereavement hallucination. By a curious irony, Mitchell himself suffered a bereavement hallucination following the sudden death of a very close friend, as Jerome Schneck described in a 1989 article:

> A reporter brought the unexpected news one morning and Mitchell, greatly shaken, went up to tell his wife. On the way back downstairs he had an odd experience: he could see the face of Brooks, larger than life, smiling, and very distinct, yet looking as if it were made of dewy gossamer. When he looked

down, the vision disappeared, but for ten days he could see it a
little above his head to the left.

Bereavement hallucinations, deeply tied to emotional needs
and feelings, tend to be unforgettable, as Elinor S., a sculptor
and printmaker, wrote to me:

> When I was fourteen years old, my parents, brother and I were
> spending the summer at my grandparents' house as we had
> done for many previous years. My grandfather had died the
> winter before.
>
> We were in the kitchen, my grandmother was at the sink,
> my mother was helping and I was still finishing dinner at
> the kitchen table, facing the back porch door. My grandfather
> walked in and I was so happy to see him that I got up to meet
> him. I said, "Grampa," and as I moved towards him, he suddenly
> wasn't there. My grandmother was visibly upset, and I thought
> she might have been angry with me because of her expression.
> I said to my mother that I had really seen him clearly, and she
> said that I had seen him because I wanted to. I hadn't been
> consciously thinking of him and still do not understand how I
> could have seen him so clearly.
>
> I am now seventy-six years of age and still remember the
> incident and have never experienced anything similar.

Elizabeth J. wrote to me about a grief hallucination experienced by her young son:

> My husband died thirty years ago after a long illness. My son
> was nine years old at the time; he and his dad ran together on a
> regular basis. A few months after my husband's death, my son

came to me and said that he sometimes saw his father running past our home in his yellow running shorts (his usual running attire). At the time, we were in family grief counselling, and when I described my son's experience, the counsellor did attribute the hallucinations to a neurologic response to grief. This was comforting to us, and I still have the yellow running shorts.

A general practitioner in Wales, W. D. Rees, interviewed nearly three hundred recently bereft people and found that almost half of them had had illusions or full-fledged hallucinations of a dead spouse. These could be visual, auditory, or both—some of the people interviewed enjoyed conversations with their hallucinated spouses. The likelihood of such hallucinations increased with the length of marriage, and they might persist for months or even years. Rees considered these hallucinations to be normal and even helpful in the mourning process.

For Susan M., bereavement stimulated a particularly vivid, multisensory experience a few hours after her mother died: "I heard the squeaking of the wheels of her walker in the hallway. She walked into the room shortly afterward and sat down on the bed next to me. I could feel her sit down on the mattress. I spoke to her and said I thought she had died. I don't remember exactly what she said in return—something about checking in with me. All I know is I could feel her there and it was frightening but also comforting."

Ray P. wrote to me after his father died at the age of eighty-five, following a heart operation. Although Ray had rushed to the hospital, his father had already lapsed into a coma. An hour before his father died, Ray whispered to him: "Dad, it's

Ray. I'll take care of mom. Don't worry, everything is going to be alright." A few nights later, Ray wrote, he was awakened by an apparition:

> I awoke in the night. I did not feel groggy or disoriented and my thoughts and vision were clear. I saw someone sitting on the corner of my bed. It was my Dad, wearing his khaki slacks and tan polo shirt. I was lucid enough to wonder initially if this could be a dream but I was certainly awake. He was opaque, not ethereal in any way, the nighttime Baltimore light pollution in the window behind him did not show through. He sat there for a moment and then said—did he speak or just convey the thought?—"Everything is all right."
>
> I turned and swung my feet to the floor. When I looked [back toward] him, he was gone. I stood and went to the bathroom, got a drink of water, and went back to bed. My dad never returned. I do not know whether this was a hallucination or something else, but since I provisionally do not believe in the paranormal, it must have been.[2]

---

2. Losing a spouse, of course, is one of the most stressful of life events, but bereavement may happen in many other situations, from the loss of a job to the loss of a beloved pet. A friend of mine was very upset when her twenty-year-old cat died, and for months she "saw" the cat and its characteristic movements in the folds of the curtains.

Another friend, Malonnie K., described a different sort of cat hallucination, after her beloved seventeen-year-old pet died:

> Much to my surprise, the next day I was getting ready for work and she appeared at the bathroom door, smiled and meowed her usual "good morning." I was flabbergasted. I went to tell my husband and when I returned, of course, she was no longer there. This was upsetting to me because I have no history of hallucinations and thought I was "above" such things. However, I have accepted that this experience was, perhaps, a result of the phenomenally close bond that we had developed and sustained over nearly two decades. I must say, I am so grateful that she stopped by one last time.

The hallucinations of grief may sometimes take a less benign form. Christopher Baethge, a psychiatrist, has written about two mothers who lost young children in a particularly traumatic way. Both had multisensory hallucinations of their dead daughters—seeing them, hearing them, smelling them, being touched by them. And both were driven to delusional, otherworldly explanations of their hallucinations: one believed that "this was her daughter's attempt to establish contact with her from another world, a world in which her daughter continues to exist"; the other heard her daughter cry out, "Mamma, don't be afraid, I'll come back."[3]

Recently I tripped over a box of books in my office, fell headlong, and broke a hip. This seemed to happen in slow motion. I thought, *I have plenty of time to put out my arm to break the fall*, but then—suddenly—I was on the floor, and as I hit, I felt the crunch in my hip. With near-hallucinatory vividness, in the next few weeks, I reexperienced my fall; it replayed itself in my mind and body. For two months I avoided the office, the place where I had fallen, because it provoked this quasi-hallucination of falling and the crunch of breaking bone. This is one example—a trivial one, perhaps—of a reaction to trauma, a mild traumatic stress syndrome. It is largely

---

3. Loss, longing, and nostalgia for lost worlds are also potent inducers of hallucinations. Franco Magnani, "the memory artist" I described in *An Anthropologist on Mars*, had been exiled from Pontito, the little village where he grew up, and although he had not returned to it in decades, he was haunted by continual dreams and hallucinations of Pontito—an idealized, timeless Pontito, as it looked before it was invaded by the Nazis in 1943. He devoted his life to objectifying these hallucinations in hundreds of nostalgic, beautiful, and uncannily accurate paintings.

resolved now, but it will, I suspect, lurk in the depths as a traumatic memory that may be reactivated under certain conditions for the rest of my life.

Much deeper trauma and consequent PTSD (post-traumatic stress disorder) may affect anyone who has lived through a violent crash, a natural cataclysm, war, rape, abuse, torture, or abandonment—any experience that produces a terrifying fear for one's own safety or that of others.

All of these situations can produce immediate reactions, but there may also be, sometimes years later, post-traumatic syndromes of a malignant and often persistent sort. It is characteristic of these syndromes that, in addition to anxiety, heightened startle reactions, depression, and autonomic disorders, there is a strong tendency to obsessive rumination on the horrors which were experienced—and, not infrequently, sudden flashbacks in which the original trauma may be reexperienced in its totality with every sensory modality and with every emotion that was felt at the time.[4] These flashbacks, though often spontaneous, are especially liable to be evoked by objects, sounds, or smells associated with the original trauma.

The term "flashback" may not do justice to the profound and sometimes dangerous delusional states that can go with post-traumatic hallucinations. In such states, all sense of the present may be lost or misinterpreted in terms of hallucination and delusion. Thus the traumatized veteran, during a flashback, may be convinced that people in a supermarket

---

4. Though "flashback" is a visual, cinematic term, auditory hallucinations can be very striking, too. Veterans with PTSD may hallucinate the voices of dying comrades, enemy soldiers, or civilians. Holmes and Tinnin, in one study, found that the hearing of intrusive voices, explicitly or implicitly accusing, affected more than 65 percent of veterans with combat PTSD.

are enemy soldiers and—if he is armed—open fire on them. This extreme state of consciousness is rare but potentially deadly.

One woman wrote to me that, having been molested as a three-year-old and assaulted at the age of nineteen, "for both events smell will bring back strong flashbacks." She continued:

> I had my first flashback of being assaulted as a child when a man sat next to me on a bus. Once I smelled [his] sweat and body odor, I was not on that bus anymore. I was in my neighbor's garage and I remembered everything. The bus driver had to ask me to get off the bus when we arrived at our destination. I lost all sense of time and place.

Particularly severe and long-lasting stress reactions may occur after rape or sexual assault. In a case reported by Terry Heins and his colleagues, for example, a fifty-five-year-old woman who had been forced to watch her parents' sexual intercourse as a young child and then forced to have intercourse with her father at the age of eight experienced repeated flashbacks of the trauma as an adult, as well as "voices"—a post-traumatic stress syndrome that was misdiagnosed as schizophrenia and led to psychiatric hospitalization.

People with PTSD are also prone to recurrent dreams or nightmares, often incorporating literal or somewhat disguised repetitions of the traumatic experiences. Paul Chodoff, a psychiatrist writing in 1963 about the effects of trauma in concentration camp survivors, saw such dreams as a hallmark of the syndrome and noted that in a surprising number of cases, they were still occurring a decade and a half after the war.[5] The same is true of flashbacks.

Chodoff observed that obsessive rumination on concentration camp experiences might diminish in some people with the passage of time, but others

> communicated an uncanny feeling that nothing of real significance had happened in their lives since their liberation, as they reported their experiences with a vivid immediacy and wealth of detail which almost made the walls of my office disappear, to be replaced by the bleak vistas of Auschwitz or Buchenwald.

Ruth Jaffe, in a 1968 article, described one concentration camp survivor who had frequent attacks in which she relived her experience at the gates of Auschwitz, where she saw her sister led off into a group destined for death but could do nothing to save her, even though she tried to sacrifice herself instead. In her attacks, she saw people entering the gates of the camp and heard her sister's voice calling, "Katy, where are you? Why do you leave me?" Other survivors are haunted by olfactory flashbacks, suddenly smelling the gas ovens—a smell which, more than anything else, brings back the horror of the camps. Similarly, the smell of burning rubble lingered around the World Trade Center for months after 9/11—and continued as

---

5. Sometimes this effect can be heightened by medications. In 1970, I had one patient with postencephalitic parkinsonism who was a concentration camp survivor. For her, treatment with L-dopa caused an intolerable exacerbation of her traumatic nightmares and flashbacks, and we had to discontinue the drug.

a hallucination to haunt some survivors even when the actual smell was gone.

There is a large body of literature on both acute stress reactions and delayed ones following natural disasters like tsunamis or earthquakes. (These occur in very young children too, though they may tend to reenact rather than hallucinate or reexperience the disaster.) But PTSD seems to have an even higher prevalence and greater severity following violence or disaster that is man-made; natural disasters, "acts of God," seem somehow easier to accept. This is the case with acute stress reactions, too: I see it often with my patients in hospital, who can show extraordinary courage and calmness in facing the most dreadful diseases but fly into a rage if a nurse is late with a bedpan or a medication. The amorality of nature is accepted, whether it takes the form of a monsoon, an elephant in musth, or a disease; but being subjected helplessly to the will of others is not, for human behavior always carries (or is felt to carry) a moral charge.

Following the First World War, some physicians felt that there must be an organic brain disturbance underlying what were then called war neuroses, which seemed unlike "normal" neuroses in many ways.[6] The term "shell shock" was coined with the notion that the brains of these soldiers had been mechanically deranged by the repeated concussion of the new high-explosive shells introduced in this war. There

---

6. In the "normal" neuroses commonly brought to psychotherapists, the buried, pathogenic material typically comes from much earlier in life. Such patients are also haunted, but as in the title of Leonard Shengold's book, they are *Haunted by Parents*.

was as yet no formal recognition of the delayed effects of the severe trauma of soldiers who endured shells and mustard gas for days on end, in muddy trenches that were filled with the rotting corpses of their comrades.[7]

Recent work by Bennet Omalu and others has shown that repeated concussion (even "mild" concussions that do not cause a loss of consciousness) can result in a chronic traumatic encephalopathy, causing memory and cognitive impairment; this may well exacerbate tendencies to depression, flash-backs, hallucination, and psychosis. Such chronic traumatic encephalopathy, along with the psychological trauma of war and injury, has been linked to the rising incidence of suicide among veterans.

That there may be biological as well as psychological determinants of PTSD would not have surprised Freud—and the treatment of these conditions may require medication as well as psychotherapy. In its worst forms, though, PTSD can be a nearly intractable disorder.

The concept of dissociation would seem crucial not only to understanding conditions like hysteria or multiple personality disorder but also to the understanding of post-traumatic syndromes. There may be an instant distancing or dissociation when a life-threatening situation occurs, as when a driver about to crash sees his car from a distance, almost like a spec-

---

7. Freud was deeply puzzled and troubled by the pertinacity of such post-traumatic syndromes after World War I. Indeed, they forced him to question his theory of the pleasure principle and, at least in this case, to see instead a much grimmer principle at work, that of repetition-compulsion, even though this seemed maladaptive, the very antithesis of a healing process.

tacle in a theater, with a sense of being a spectator rather than a participant. But the dissociations of PTSD are of a more radical kind, for the unbearable sights, sounds, smells, and emotions of the hideous experience get locked away in a separate, subterranean chamber of the mind.

Imagination is qualitatively different from hallucination. The visions of artists and scientists, the fantasies and daydreams we all have, are located in the imaginative space of our own minds, our own private theaters. They do not normally appear in external space, like the objects of perception. Something has to happen in the mind/brain for imagination to overleap its boundaries and be replaced by hallucination. Some dissociation or disconnection must occur, some breakdown of the mechanisms that normally allow us to recognize and take responsibility for our own thoughts and imaginings, to see them as ours and not as external in origin.

It is not clear, however, that such a dissociation can explain everything, for quite different sorts of memory may be involved. Chris Brewin and his colleagues have argued that there is a fundamental difference between the extraordinary flashback memories of PTSD and those of ordinary autobiographic memory and have provided much psychological evidence for such a difference. Brewin et al. see a radical distinction between autobiographic memories, which are verbally accessible, and flashback memories, which are not verbally or voluntarily accessible but may erupt automatically if there is any reference to the traumatic event or something (a sight, a smell, a sound) associated with it. Autobiographic memories are not isolated—they are embedded in the context of an entire life, given a broad and deep context and perspective—and they can be revised in relation to different contexts and perspectives. This is not the case with traumatic memories. The survivors of trauma may be unable to achieve the detachment of retrospection or recollection; for them the traumatic events, in all their fearfulness and horror, all their sensorimotor vividness and concreteness, are sequestered. The events seem to be preserved in a different form of memory, isolated and unintegrated.

Given this isolation of traumatic memory, the thrust of psychotherapy must be to release the traumatic events into the light of full consciousness, to reintegrate them with autobiographic memory. This can be an exceedingly difficult and sometimes nearly impossible task.

The idea that different sorts of memory are involved gets strong support from the survivors of traumatic situations who do not get PTSD and are able to live full, unhaunted lives. One such person is my friend Ben Helfgott, who was incarcerated in a concentration camp between the ages of twelve and sixteen. Helfgott has always been able to talk fully and freely about his experiences during these years, about the killing of his parents and family and the many horrors of the camps. He can recall it all in conscious, autobiographic memory; it is an accepted, integrated part of his life. His experiences were not locked away as traumatic memories, but he knows the other side well—he has seen it in hundreds of others: "The ones who 'forget,'" he says, "they suffer later." Helfgott is one of the contributors to *The Boys*, a remarkable book by Martin Gilbert that relates the stories of hundreds of boys and girls who, like Helfgott, survived years in concentration camps but somehow emerged relatively undamaged and have never been subject to PTSD or hallucinations.

A deeply superstitious and delusional atmosphere can also foster hallucinations arising from extreme emotional states, and these can affect entire communities. In his 1896 Lowell Lectures (collected as *William James on Exceptional Mental States*) James included lectures on "demoniacal possession" and witchcraft. We have very detailed descriptions of the hallucinations characteristic of both states—hallucinations which rose, at times, to epidemic proportions and were ascribed to the workings of the devil or his minions, but which we can now interpret as the effects of suggestion and even torture in societies where religion had taken on a fanatical character. In his book *The Devils of Loudun*, Aldous Huxley described the delusions of demonic possession that swept over the French village of Loudun in 1634, starting with a mother superior and all the nuns in an Ursuline convent. What began as Sister Jeanne's religious obsessions were magnified to a state of hallucination and hysteria, in part by the exorcists themselves, who, in effect, confirmed the entire community's fear of demons. Some of the exorcists were affected as well. Father Surin, who had been closeted for hundreds of hours with Sister Jeanne, was himself to be haunted by religious hallucinations of a terrifying nature. The madness consumed the entire village, just as it would later do in the infamous Salem witch trials.[8]

---

8. Many of the testimonies and accusations in the Salem witch trials described assaults by hags, demons, witches, or cats (which were seen as witches' familiars). The cats would sit astride sleepers, pressing on their chests, suffocating them, while the sleepers had no power to move or resist. These are experiences we would now interpret in terms of sleep paralysis and night-mare, but which were given a supernatural narrative. The whole subject is explored by Owen Davies in his 2003 article "The Nightmare Experience, Sleep Paralysis, and Witchcraft Accusations."

Other conditions have also been suggested as contributing to the

The conditions and pressures in Loudun or Salem may have been extraordinary, though witch-hunting and forced confession have hardly vanished from the world; they have simply taken other forms.

Severe stress accompanied by inner conflicts can readily induce in some people a splitting of consciousness, with varied sensory and motor symptoms, including hallucinations. (The old name for this condition was hysteria; it is now called conversion disorder.) This seemed to be the case with Anna O., the remarkable patient described by Freud and Breuer in their *Studies on Hysteria.* Anna had little outlet for her intellectual or sexual energies and was strongly prone to daydreaming—she called it her "private theater"—even before her father's final illness and death pushed her into a splitting or dissociation of personality, an alternation between two states of consciousness. It was in her "trance" state (which Breuer and Freud called an "auto-hypnotic" state) that she had vivid and almost always frightening hallucinations. Most commonly she would

hallucinations and hysteria of seventeenth-century New England. One hypothesis, which Laurie Winn Carlson proposes in her book *A Fever in Salem,* sees the madness as a manifestation of a postencephalitic disorder.

Others have proposed that ergot poisoning played a part. Ergot, a fungus containing toxic alkaloid compounds similar to LSD, can infest rye and other grains, and if contaminated bread or flour is eaten, ergotism may result. This happened frequently in the Middle Ages, and it could cause agonizing gangrene (which led to one of its popular names, St. Anthony's fire). Ergotism could also cause convulsions and hallucinations very similar to those of LSD.

In 1951, an entire French village succumbed to ergot poisoning, as John Grant Fuller described in his book *The Day of St. Anthony's Fire.* Those affected endured several weeks of terrifying hallucinations and often compulsions to jump from windows, as well as extreme insomnia.

see snakes, her own hair as snakes, or her father's face transformed into a death's-head. She retained no memory or consciousness of these hallucinations until she was again in a
hypnotic trance, but this time induced by Breuer:

> She used to hallucinate in the middle of a conversation, run
> off, start climbing up a tree, etc. If one caught hold of her, she
> would very quickly take up her interrupted sentence without
> knowing anything about what had happened in the interval.
> All these hallucinations, however, came up and were reported
> on in her hypnosis.

Anna's "trance" personality became more and more domi-
nant as her illness progressed, and for long periods she would
be oblivious or blind to the here and now, hallucinating herself
as she was in the past. She was, at this point, living largely in
a hallucinatory, almost delusional world, like the nuns of Loudun or the "witches" of Salem.

But unlike the witches, the nuns, or the tormented survivors of concentration camps and battles, Anna O. enjoyed an
almost complete recovery from her symptoms, and went on to
lead a full and productive life.

That Anna, who was unable to remember her hallucinations
when "normal," could remember all of them when she was
hypnotized, shows the similarity of her hypnotized state to her
spontaneous trances.

Hypnotic suggestion, indeed, can be used to induce hallucinations.[9] There is, of course, a world of difference between

9. This was shown experimentally by Brady and Levitt in a 1966 study,
in which they suggested to hypnotized subjects that they "see" (i.e.,
hallucinate) a moving visual stimulus (a rotating drum with vertical

the long-lasting pathological state we call hysteria and the brief trance states which can be induced by a hypnotist (or by oneself). William James, in his lectures on exceptional mental states, referred to the trances of mediums who channel voices and images of the dead, and of scryers who see visions of the future in a crystal ball. Whether the voices and visions in these contexts were veridical was of less concern to James than the mental states which could produce them. Careful observation (he attended many séances) convinced him that mediums and crystal gazers were not usually conscious charlatans or liars in the ordinary sense; nor were they confabulators or phantasts. They were, he came to feel, in altered states of consciousness conducive to hallucinations—hallucinations whose content was shaped by the questions they were asked. These exceptional mental states, he thought, were achieved by self-hypnosis (no doubt facilitated by poorly lit and ambiguous surroundings and the eager expectations of their clients).

Such practices as meditation, spiritual exercises, and ecstatic drumming or dancing can also facilitate the achievement of trance states akin to that of hypnosis, with vivid hallucinations and profound physiological changes (for instance, a rigidity which allows the entire body to remain as stiff as a board while supported only at the head and feet). Meditative or contemplative techniques (often aided by sacred music, painting, or architecture) have been used in many religious traditions—sometimes to induce hallucinatory visions. Studies by

---

stripes). The subjects' eyes, as they did this, showed the same automatic tracking movements ("optokinetic nystagmus") that occur when one is actually looking at such a rotating drum—whereas no such movements occur (and they are impossible to feign) if one merely imagines such a visual target.

Andrew Newberg and others have shown that long-term practice of meditation produces significant alterations in cerebral
blood flow in parts of the brain related to attention, emotion,
and some autonomic functions.

The commonest, the most sought, and (in many cultures
and communities) the most "normal" of exceptional
mental states is that of a spiritually attuned consciousness, in
which the supernatural, the divine, is experienced as material
and real. In her remarkable book *When God Talks Back*, the
ethnologist T. M. Luhrmann provides a compelling examination of this phenomenon.

Luhrmann's earlier work, on people who practice magic in
present-day Britain, involved entering their world very fully.
"I did what anthropologists do," she writes. "I participated in
their world: I joined their groups. I read their books and novels.
I practiced their techniques and performed in their rituals. For
the most part, I found, the rituals depended on techniques of
the imagination. You shut your eyes and saw with your mind's
eye the story told by the leader of the group." She was intrigued
to find that, after about a year of this practice, her own mental
imagery became clearer, more detailed, and more solid; and
her concentration states became "deeper and more sharply different from the everyday." One night she became immersed
in a book about Arthurian Britain, "giving way," she writes,
"to the story and allowing it to grip my feelings and to fill my
mind." The next morning she woke up to a striking sight:

> I saw six druids standing against the window, above the stirring
> London street below. I *saw* them, and they beckoned to me. I

stared for a moment of stunned astonishment, and then I shot up out of bed, and they were gone. Had they been there in the flesh? I thought not. But my memory of the experience is very clear. . . . I remember that I saw them as clearly and distinctly and as external to me as I saw the notebook in which I recorded the moment. I remember it so clearly because it was so singular. Nothing like that had ever happened to me before.

Later, Luhrmann embarked on a study of evangelical religion. The very essence of divinity, of God, is immaterial. God cannot be seen, felt, or heard in the ordinary way. How, she wondered, in the face of this lack of evidence, does God become a real, intimate presence in the lives of so many evangelicals and other people of faith? Many evangelicals feel they have literally been touched by God, or heard his voice aloud; others speak of feeling his presence in a physical way, of knowing that he is there, walking beside them. The emphasis in evangelical Christianity, Luhrmann writes, is on prayer and other spiritual exercises as skills that must be learned and practiced. Such skills may come more easily to people who are prone to being completely engaged, fully absorbed, by their experiences, whether real or imaginary—the capacity, Luhrmann writes, "to focus in on the mind's object . . . the mode of the novel reader and the music listener and the Sunday hiker, caught up in imagination or appreciation." Such a capacity for absorption, she feels, can be honed with practice, and this is part of what happens in prayer. Prayer techniques are often focused on attention to sensory detail:

[Congregants] practice seeing, hearing, smelling, and touching in the mind's eye. They give these imagined experiences the

sensory vividness associated with the memories of real events.
What they are able to imagine becomes more real to them.

And one day the mind leaps from imagination to hallucination, and the congregant *hears* God, *sees* God.

These yearned-for voices and visions have the reality of perception. One of Luhrmann's subjects, Sarah, put it this way: "The images I see [in prayer] are very real and lucid. Different from just daydreaming. I mean, sometimes it's almost like a PowerPoint presentation." Over time, Luhrmann writes, Sarah's images "got richer and more complicated. They seemed to have sharper borders. They continued to get more complex and more distinct." Mental images become as clear and as real as the external world.

Sarah had many such experiences; some congregants might have only a single one—but even a single experience of God, imbued with the overwhelming force of actual perception, can be enough to sustain a lifetime of faith.

Even at a more modest level, all of us are susceptible to the power of suggestion, especially if it is combined with emotional arousal and ambiguous stimuli. The idea that a house is "haunted," though scoffed at by the rational mind, may nonetheless induce a watchful state of mind and even hallucination, as Leslie D. brought out in a letter to me:

> Almost four years ago I started a job that is housed in one of the oldest residences in Hanover, PA. On my first day, I was told there was a resident ghost, the ghost of Mr. Gobrecht, who lived here many years ago and was a music teacher. . . .

I suppose he died in the house. It would be almost impossible to adequately describe how much I do NOT believe in the supernatural! However, within days I started to feel something like a hand tugging on my pant leg while I sat at my desk, and once in a while a hand on my shoulder. Just a week ago we were discussing the ghost, and I felt (*very pronounced*) fingers moving along my upper back, just behind my shoulder, distinct enough to make me jump. Power of suggestion, maybe?

Children not uncommonly have imaginary companions. Sometimes this may be a sort of ongoing, systematized daydreaming or storytelling, the creation of an imaginative and perhaps lonely child; in some cases it may have elements of hallucination—a hallucination that is benign and pleasant, as Hailey W. described to me:

Growing up without brothers or sisters, I created a few imaginary friends whom I played with frequently from approximately age three to six. The most memorable of these was a pair of identical twin girls named Kacey and Klacey. They were my age and size, and we would often do things together like play on the swings in the backyard or have tea parties. Kacey and Klacey also had a little sister named Milky. I had a strong image of them all in my mind's eye, and they seemed very real to me at the time. My parents were mostly amused by it, though they did question whether it was natural for my imaginary friends to be so detailed and plentiful. They recall me having long conversations at the table with "no one," and when asked, I would always say I was talking with Kacey and Klacey. Often when playing (with toys, or games) I would say I was playing with Kacey and Klacey or

Milky. I would talk about them often as well, and for a period of time I remember being fixated on the idea of a seeing-eye dog, begging my mother to let me have one. Rather taken aback, my mother asked where I got the idea; I replied that Kacey and Klacey's mother was blind, and that I wanted a seeing-eye dog like hers. As an adult, I am still surprised when someone tells me that they never had imaginary friends growing up, as they were such an important—and enjoyable—part of my childhood.

And yet "imagination" may not be an adequate term here, for imaginary companions may seem intensely real, as no other products of fantasy or imagination do. Perhaps the difficulty of fitting our adult categories of "reality" and "imagination" to the thoughts and play of children is not surprising; for, if Piaget is right, children cannot consistently and confidently distinguish fantasy from reality, inner from outer worlds, until the age of seven or so. It is usually at this age, or a little later, that imaginary companions tend to disappear.

Children may also be more accepting of their hallucinations, having not yet learned that hallucinations are considered (in our culture) "abnormal." Tom W. wrote to me about his "intended" childhood hallucinations, hypnagogic visions he would bring on as entertainment from the ages of four to seven:

> I used to entertain myself while falling asleep by hallucinating. I would lie in bed and stare up at the ceiling in the half-light. . . . I would stare at a fixed point, and by holding my eyes very still, the ceiling would neutralize and gradually become swarming pixels, which would become patterns: waves and

grids and paisleys. Then, in the midst of that, figures would start to appear and interact. I remember quite a few—[and] I remember the exceptional visual clarity of them. Once the vision was present, I could look around at things the way you would a film.

There was another way I used to do this. There was a family portrait that hung at the foot of my bed, a classical staged photo of my grandparents, cousins, an aunt and uncle, my parents, my brother, and me. Behind us was a huge privet hedge. Again, in the evening, I would gaze at the portrait. Very quickly, strange and delightfully silly things would start to happen: apples would grow out of the privet hedge, my cousins would begin to chatter and chase each other around the group. My grandmother's head would "pop off" and attach to her two calves, which would then start to dance about. Grim as that seems now, I found it hilarious then.

At the other end of life, there is a special sort of hallucination that may attend death or the anticipation of death. Working in old-age homes and nursing homes, I have been struck and moved by how often patients who are lucid, sane, and fully conscious may have hallucinations when they feel that death is near.

When Rosalie—the very old blind lady I described in the chapter on Charles Bonnet syndrome—became ill and thought she was dying, she had visions of her mother and heard her mother's voice welcoming her into heaven. These hallucinations were completely different in character from her usual CBS hallucinations—they were multisensory, personal, addressed to her, and steeped in warmth and tenderness. Her CBS hallucinations, by contrast, had no apparent relation to her and aroused no emotion. I have known other patients (who did not have CBS or any other special condition facilitating hallucinations) to have similar deathbed hallucinations—sometimes the first and last hallucinations in their lives.

## 14

# Doppelgängers:
# Hallucinating Oneself

Sleep paralysis may be associated, as some of my corre-spondents have emphasized, with a sense of floating or levitation, and even hallucinations of leaving one's own body and flying through space. These experiences, so unlike the hideous night-mare ones, may go with feelings of calm and joy (some of Cheyne's subjects used the term "bliss"). Jeanette B., who has had a lifetime of narcolepsy and sleep paralysis (which she refers to as "spells"), described this to me:

> It was after college that the spells became both a burden and a blessing. Not being able to pull myself out of the paralysis one night, I let go; and felt myself slowly rising out of my body! I had come through the terror part and felt a wonderful peaceful bliss as I rose out of my body and floated up. Now, as I experienced this I found it very difficult to believe it was a hallucination. All of my senses seemed unusually sharp: someone's radio playing in another room, crickets chirping outside the window.

Without going into detail, this was a hallucination that was more pleasant than anything I had ever experienced. . . .

I suppose I became so nearly addicted to the out of body experiences, that when offered some meds from my neurologist to help with the nighttime paralysis and hallucinations, I refused, rather than give up the out of body experiences. I didn't say that was the reason.

For quite some time, I would try to will myself into that pleasant hallucination. I discovered it usually came after much stress or lack of sleep, and would deprive myself of sleep in order to achieve the experience of floating amongst the stars, high enough to observe the curvature of the earth. . . .

But bliss can coincide with terror—Peter S., a friend, found this when he had a single episode of sleep paralysis with hallucination. It seemed to him that he left his body, cast a backward glance at it, then soared up into the sky. He had an enormous sense of freedom and joy, now that he had left the limitations of his human body, a feeling that he could roam at will through the universe. But there was also a fear, which became terror, that he might be lost forever in infinity, unable to rejoin his body on earth.

Out-of-body experiences may occur when specific regions of the brain are stimulated in the course of a seizure or a migraine, as well as with electrical stimulation of the cortex.[1] They may

---

1. The term "out-of-body experience" was introduced in the 1960s by Celia Green, an Oxford psychologist. While there had been stories of out-of-body experiences for centuries, Green was the first to systematically examine a large number of firsthand accounts, from more than four hundred people whom she located by launching a public appeal through the newspapers and the BBC. In her 1968 book, *Out-of-the-Body Experiences*, she analyzed these in detail.

occur with drug experiences and in self-induced trances. OBEs can also occur when the brain is not receiving enough blood, as may happen if there is a cardiac arrest or arrhythmia, massive blood loss, or shock.

My friend Sarah B. had an OBE in the delivery room, just after giving birth. She had delivered a healthy baby, but she had lost a lot of blood, and her obstetrician said that he would have to compress the uterus to stop the hemorrhage. Sarah wrote:

> I felt my uterus being squeezed and told myself not to move or cry out. . . . Then, suddenly, I was floating with the back of my head against the ceiling. I was looking down on a body which was not my own. The body was some distance from me. . . . I watched the doctor pound on this woman and heard him grunt loudly with his efforts. I thought, "This woman is very inconsiderate. She is giving Dr. J. a lot of trouble." . . . So I was completely oriented to time, day, place, people, and event. I was just unaware that the center of the drama was myself.
>
> After some time, Dr. J. withdrew his hands from the body, stepped back, and announced that the bleeding had stopped. As he said this, I felt myself slip back into my body like an arm sliding into the sleeve of a coat. I was no longer looking, from a distance, down on the doctor; instead he was looming above and quite close to me. His green surgical scrubs were covered with blood.

Sarah had critically low blood pressure, and it was probably this—her brain getting insufficient oxygen—which precipitated the OBE. Anxiety may have constituted an additional factor, as reassurance did in ending the attack, despite her still very low blood pressure. Her not recognizing her own body is curious, though it is commonly reported that the body looks "vacated" or "empty" when the now-disembodied self looks down on its former home.

Another friend, Hazel R., a chemist, told me of an experience she had many years ago, also when she was in labor. She was offered heroin for her pain (this was common in England in those days), and as the heroin took effect, she felt herself floating upwards, coming to rest beneath the ceiling in the corner of the delivery room. She saw her body beneath her, and she had no pain whatever—she felt that the pain had stayed in the body below her. She also had a sense of great visual and intellectual acuity: she felt that she could easily solve any problem (unfortunately, she said wryly, no problem presented itself). As the heroin wore off, she returned to her body and its violent contractions and pain. When her obstetrician told her she could have a further dose, she asked if it could affect the baby adversely. Once she was reassured that it would not, she assented to a second dose, and again she enjoyed a detachment from her body and its labor pains, as well as a feeling of supernal mental clarity.[2] Although this occurred more than fifty years ago, Hazel still remembers every detail.

It is not easy to imagine such detachment from the body if one has never experienced it. I have never had an OBE myself, but I was once involved in a remarkably simple experiment which showed me how easily one's sense of self can be detached from one's body and "reembodied" in a robot. The robot was

---

2. Several of Celia Green's subjects described similar feelings. "My mind was clearer and more active than before," one wrote; another spoke of being "all-knowing and understanding." Green wrote that such subjects felt they "could obtain an answer to any question they chose to formulate."

a massive metal figure with video cameras for "eyes" and lob-sterlike claws for "hands," designed for training astronauts to operate similar machines in space. I donned goggles connected to the video cameras, so that in effect I was seeing the world through the robot's eyes, and I inserted my hands into gloves with sensors that would register my movements and transmit them to the robot's claws. As soon as I was connected, look-ing out through the robot's eyes, I had the odd experience of seeing, a few feet to my left, an oddly small figure (did it seem small because I, embodied in the robot, was now so large?) sit-ting in a chair and wearing goggles and gloves, a vacated figure who I realized, with a start, must be *me*.

Tony Cicoria, a surgeon, was struck by lightning a few years ago and suffered a cardiac arrest. (I tell his whole, complex story in *Musicophilia*.) He recounted this to me:

> I remember a flash of light . . . hit me in the face. Next thing I remember, I was flying backwards . . . [then] I was flying forwards. I saw my own body on the ground. I said to myself, "Oh shit, I'm dead." I saw people converging on the body. I saw a woman . . . position herself over my body, give it CPR.

Cicoria's OBE became more complex. "There was a bluish-white light . . . an enormous feeling of well-being and grace"; he felt he was being drawn into heaven (his OBE had evolved into a "near-death experience," which is not the case with most OBEs), and then—it could have been little more than thirty or forty seconds from the moment he was struck by lightning—"Slam! I was back."

The term "near-death experience" (NDE) was introduced by Raymond Moody in his 1975 book *Life After Life.* Moody, culling information from many interviewees, delineated a remarkably uniform and stereotyped set of experiences common to many NDEs. A majority of people felt that they were being drawn into a dark tunnel and then propelled towards a brightness (which some interviewees called "a being of light"); and, finally, they sensed a limit or barrier ahead—most interpreted this as the boundary between life and death. Some experienced a rapid replay or review of events in their lives; others saw friends and relatives. In a typical NDE, all this was suffused with a sense of peace and joy so intense that being "forced back" (into one's body, into life) might be accompanied by a strong sense of regret. Such experiences were felt as real—"more real than real," as was often commented. Many of Moody's interviewees favored a supernatural interpretation for these remarkable experiences, but others have increasingly tended to regard them as hallucinations, albeit of an extraordinarily complex kind. A number of researchers have sought a natural explanation in terms of brain activity and blood flow, since NDEs are especially associated with cardiac arrest and may also occur in faints, when blood pressure plunges, the face becomes ashen, and the head and brain are drained of blood.

Kevin Nelson and his colleagues at the University of Kentucky have presented evidence suggesting that, with the compromise of cerebral blood flow, there is a dissociation of consciousness so that, although awake, the subjects are paralyzed and subject to the dreamlike hallucinations characteristic of REM sleep ("REM intrusions")—in a state, therefore, with resemblances to sleep paralysis (NDEs are also commoner in people prone to sleep paralysis). Added to this are various special features: the "dark tunnel" is correlated, Nelson feels, with the compromise of blood flow to the retinas (this is well known to produce a constriction of the visual fields, or tunnel vision, and may occur in pilots subjected to high g-stresses). The "bright light" Nelson correlates with a flow of neuronal excitement moving from a part of the brain stem (the pons) to subcortical visual relay stations and then to the occipital cortex. Added to all these neurophysiological changes may be a sense of terror and awe going with the knowledge that one is undergoing a mortal crisis—some subjects have actually heard themselves pronounced dead—and the wish that dying, if imminent and inevitable, should be peaceful and perhaps a passage to a life after death.

Both Olaf Blanke and Peter Brugger have studied such phenomena in several patients with severe epilepsy. Like Wilder Penfield's patients in the 1950s, people with intractable seizures that do not respond to medication may need surgery to remove the epileptic focus responsible. Such surgery requires extensive testing and mapping to find the seizure focus and to avoid damaging vital areas. The patient must be awake during this procedure, so that he can report what he is experiencing. Blanke was able to demonstrate that stimulating certain areas of the brain's right angular gyrus invariably caused OBEs in one such patient, as well as feelings of lightness and levitation and changes in body image; the patient saw her legs "becoming shorter" and moving towards her face. Blanke et al. speculate that the angular gyrus is a crucial node in a circuit that mediates body image and vestibular sensations, and that "the experience of dissociation of self from the body is a result of failure to integrate information from the body with vestibular information."

At other times, one is not disembodied but sees a double of oneself from one's normal viewpoint, and the other self often mimics (or shares) one's own postures and movements. These autoscopic hallucinations are purely visual and usually fairly brief—they may occur, for instance, in the few minutes of a migraine or epileptic aura. In his delightful history of migraine, "Migraine: From Cappadocia to Queen Square," Macdonald Critchley describes this in the great naturalist Carl Linnaeus:

> Often Linnaeus saw "his other self" strolling in the garden parallel with himself, and the phantom would mimic his movements, i.e. stoop to examine a plant or to pick a flower. Sometimes the alter ego would occupy his own seat at his library desk. Once at a demonstration to his students he wanted to fetch a specimen from his room. He opened the door rapidly, intending to enter, but pulled up at once saying, "Oh! I'm there already."

A similar hallucination of a double was seen regularly by Charles Lullin, the grandfather of Charles Bonnet, for about three months, as Douwe Draaisma describes:

> One morning as he was quietly smoking his pipe at the window, he saw on his left a man leaning casually against the window frame. Except for the fact that he was a head taller, the man looked exactly like him: he was also smoking a pipe, and he was wearing the same cap and the same dressing gown. The man was there again the next morning, and he gradually became a familiar apparition.

The autoscopic double is literally a mirror image of oneself, with right transposed to left and vice versa, mirroring one's positions and actions. The double is a purely visual phenomenon, with no identity or intentionality of its own. It has no desires and takes no initiatives; it is passive and neutral.[3]

Jean Lhermitte, reviewing the subject of autoscopy in 1951, wrote: "The phenomenon of the double can be produced by many other diseases of the brain besides epilepsy. It appears in general paralysis [neurosyphilis], in encephalitis, in encephalosis of schizophrenia, in focal lesions of the brain, in post-traumatic disorders. . . . The apparition of the double should make one seriously suspect the incidence of a disease."

It is now thought that a substantial number—perhaps a third—of all cases of autoscopy may be associated with schizophrenia, and even cases of manifestly physical or organic origin may be sensitive to suggestion. T. R. Dening and German Berrios described a thirty-five-year-old man whose apparitions

---

3. August Strindberg noted, in his autobiographical novel *Inferno*, an odd body double, an "other" who mirrored his every movement.

> This unknown man never uttered a word; he seemed to be occupied in writing something behind the wooden partition that separated us. All the same, it was odd that he should push back his chair every time I moved mine. He repeated my every movement in a way that suggested that he wanted to annoy me by imitating me. . . . When I went to bed the man in the room next to my desk went to bed too. . . . I could hear him lying there, stretched out parallel to me. I could hear him turning the pages of a book, putting out the lamp, breathing deeply, turning over and falling asleep.

Strindberg's "unknown man" is identical with Strindberg in one sense: a projection of him, at least of his movements, his actions, his body image. Yet, at the same, he is someone else, an Other who occasionally "annoys" Strindberg, but perhaps, at other times, seeks to be companionable. He is, in the literal sense of the term, Strindberg's "Other," his "alter ego."

were related to temporal lobe seizures following a head injury. The man said that he had once seen his ties hanging up as a rack of snakes, but when asked whether he had any outright hallucinations or autoscopic experiences, he said no. A week later, he came in for another appointment in a state of some excitement, for he had now had an autoscopic experience:

> He had been sitting in a café, when he was suddenly aware of an image of himself, about 15–20 yards away, looking in through the café window. The image was dark and looked like him at the age of nineteen (when his accident occurred). It did not speak and probably lasted for less than a minute. He felt amazed and uncomfortable, as though physically struck, and he felt he had to get up and leave. It is difficult not to suppose that the timing of this episode was influenced by the questions asked by the psychiatrist in the previous week.

While most examples of autoscopy are fairly brief, long-lasting autoscopy has also been recorded. Zamboni et al. provide a detailed description of this in a 2005 paper. Their patient, B.F., was a young woman who developed eclampsia in pregnancy and was comatose for two days. As she started to recover, it was evident that she was cortically blind and had a partial paralysis on both sides as well as an unawareness of her left side and of the left side of space, a hemi-neglect. With further recovery, her visual fields became full and she could discriminate color, but she was profoundly agnosic, unable to recognize objects or even shapes. At this stage, Zamboni et al. wrote, their patient first started seeing her own image as if reflected in a mirror, about a meter in front of her. The image was transparent, as though it were set "in a sheet of glass," but a bit blurry. It was life-sized and consisted of a head and shoulders, though if she looked down, she could see its legs, too. It was always dressed exactly like her. It disappeared when she closed her eyes and reappeared the moment she opened them (although, as the novelty wore off, she was able to "forget" the image for hours at a time). She had no special feelings for the image and never attributed any thoughts or feelings or intentionality to it.

As B.F.'s agnosia disappeared, the mirror image gradually faded, and it had vanished entirely by six months after the original brain injury. Zamboni et al. suggested that the unusual persistence of this mirror image may have been associated with her severe visual loss, along with disturbances of multisensory integration (visual, tactile, proprioceptive, etc.) at higher levels, perhaps in the parietotemporal junction.

An even stranger and more complex form of hallucinating oneself occurs in "heautoscopy," an extremely rare form of autoscopy where there is interaction between the person and his double; the interaction is occasionally amiable but more often hostile. Moreover, there may be deep bewilderment as to who is the "original" and who the "double," for consciousness and sense of self tend to shift from one to the other. One may see the world first with one's own eyes, then through the double's eyes, and this can provoke the thought that he—the other—is the real person. The double is not construed as passively mirroring one's posture and actions, as with autoscopy; the heautoscopic double can do, within limits, whatever it wants to (or it may lie still, doing nothing at all).

"Ordinary" autoscopy—such as Linnaeus and Lullin

experienced—seems relatively benign; the hallucination is purely visual, a mirroring which appears only occasionally, has no pretensions to autonomy, no intentionality, and attempts no interactions. But the heautoscopic double, mocking or stealing one's identity, may arouse feelings of fear and horror and provoke impulsive and desperate acts. In a 1994 paper, Brugger and his colleagues described such an episode in a young man with temporal lobe epilepsy:

> The heautoscopic episode occurred shortly before admission. The patient stopped his phenytoin medication, drank several glasses of beer, stayed in bed the whole of the next day, and in the evening he was found mumbling and confused below an almost completely destroyed large bush just under the window of his room on the third floor. . . . The patient gave the following account of the episode: on the respective morning he got up with a dizzy feeling. Turning around, he saw himself still lying in bed. He became angry about "this guy who I knew was myself and who would not get up and thus risked being late at work." He tried to wake the body in the bed first by shouting at it; then by trying to shake it and then repeatedly jumping on his alter ego in the bed. The lying body showed no reaction. Only then did the patient begin to be puzzled about his double existence and become more and more scared by the fact that he could no longer tell which of the two he really was. Several times his bodily awareness switched from the one standing upright to the one still lying in bed; when in lying in bed mode he felt quite awake but completely paralysed and scared by the figure of himself bending over and beating him. His only intention was to become one person again and, looking out of the window (from where he could still see his body lying

in bed), he suddenly decided to jump out "in order to stop the intolerable feeling of being divided in two." At the same time, he hoped that "this really desperate action would frighten the one in bed and thus urge him to merge with me again." The next thing he remembers is waking up in pain in the hospital.

The term "heautoscopy" (sometimes spelled héautoscopy), introduced in 1935, is not always regarded as a useful one. T. R. Dening and German Berrios, for example, write, "We see no advantage in this term; it is pedantic, almost unpronounceable, and not widely used in ordinary practice." They see not a dichotomy but a continuum or spectrum of autoscopic phenomena, in which the sense of relationship to one's autoscopic image may vary from minimal to intense, from indifferent to impassioned, and the sense of its "reality" may be equally variable and inconsistent. In a 1955 paper, Kenneth Dewhurst and John Pearson described a schoolteacher who, at the start of a subarachnoid hemorrhage, saw an autoscopic "double" for four days:

It appeared quite solid as if seen in a mirror, dressed exactly as he was. It accompanied him everywhere; at meal-times it stood behind his chair and did not reappear till he had finished eating. At night it would undress and lie down on the table or couch in the next room of his flat. The double never said anything to him or made any sign, but only repeated his actions: it had a constant sad expression. It was obvious to the patient that this was all a hallucination, but nevertheless it had become sufficiently a part of himself for the patient to draw a chair up for his double when he first visited his private doctor.

In 1844, a century before the term was coined, A. L. Wigan, a physician, described an extreme case of heautoscopy with tragic consequences:

> I knew a very intelligent and amiable man, who had the power of thus placing before his eyes *himself,* and often laughed heartily at *his double*, who always seemed to laugh in turn. This was long a subject of amusement and joke, but the ultimate result was lamentable. He became gradually convinced that he was haunted by [his other] *self.* This other self would argue with him pertinaciously, and to his great mortification sometimes refute him, which, as he was very proud of his logical powers, humiliated him exceedingly. He was eccentric, but was never placed in confinement or subjected to the slightest restraint. At length, worn out by the annoyance, he deliberately resolved not to enter on another year of existence—paid all his debts— wrapped up in separate papers the amount of the weekly demands—waited, pistol in hand, the night of the 31st of December, and as the clock struck twelve fired it into his mouth.

The theme of the double, the doppelgänger, a being who is partly one, partly Other, is irresistible to the literary mind, and is usually portrayed as a sinister portent of death or calamity. Sometimes, as in Edgar Allan Poe's "William Wilson," the double is the visible and tangible projection of a guilty conscience that grows more and more intolerable until, finally, the victim turns murderously on his double and finds that he has stabbed himself. Sometimes the double is invisible and intangible, as in Guy de Maupassant's story "Le Horla," but this double nonetheless leaves evidence of his existence (for instance, he drinks the water that the narrator sets out in his night bottle).

At the time he wrote this, de Maupassant often saw a double himself, an autoscopic image. As he remarked to a friend, "Almost every time when I return home I see my double. I open the door and see myself sitting in the armchair. I know it's a hallucination the moment I see it. But isn't it remarkable? If you had not a cool head, wouldn't you be afraid?"

De Maupassant had neurosyphilis at this point, and when the disease grew more advanced, he became unable to recognize himself in a mirror and, it is reported, would greet his image in a mirror, bow, and try to shake hands with it.

The persecuting yet invisible Horla, while perhaps inspired by such autoscopic experiences, is a different thing altogether; it belongs, like William Wilson and Golyadkin's double in Dostoevsky's novella, to the essentially literary, Gothic genre of the doppelgänger, a genre which had its heyday from the late eighteenth century to the turn of the twentieth.

In real life—despite the extreme cases reported by Brugger and others—heautoscopic doubles may be less malign; they may even be good-natured or constructive moral figures. One of Orrin Devinsky's patients, who had heautoscopy in association with his temporal lobe seizures, described this episode: "It was like a dream, but I was awake. Suddenly, I saw myself about five feet in front of me. My double was mowing the lawn, which is what I should have been doing." This man subsequently had more than a dozen such episodes just before seizures, and many others that were apparently unrelated to seizure activity. In a 1989 paper, Devinsky et al. wrote:

His double is always a transparent, full figure that is slightly smaller than life size. It often wears different clothing than the patient and does not share the patient's thoughts or emotions. The double is usually engaged in an activity that the patient feels he should be doing, and he says, "that guy is my guilty conscience."

Embodiment seems to be the surest thing in the world, the one irrefutable fact. We think of ourselves as being in our bodies, and of our bodies as belonging to us, and us alone: thus we look out on the world with our own eyes, walk with our own legs, shake hands with our own hands. We have a sense, too, that consciousness is in our own head. It has long been assumed that the body image or body schema is a fixed and stable part of one's awareness, perhaps in part hardwired, and largely sustained and affirmed by the continuing proprioceptive feedback from joint and muscle receptors regarding the position and movement of one's limbs.

There was general astonishment, therefore, when Matthew Botvinick and Jonathan Cohen showed in 1998 that a rubber hand, under the right circumstances, could be mistaken for one's own. If a subject's real hand is hidden under a table while the rubber hand is visible before him, and both are stroked in synchrony, then the subject has the convincing illusion, even though he knows better, that the rubber hand is his—and that the sensation of being stroked is located in this inanimate though lifelike object. As I found when I looked through the "eyes" of a robot, knowledge in such a situation does nothing to dispel the illusion. The brain does its best to correlate all the senses, but the visual input here trumps the tactile.

Henrik Ehrsson, in Sweden, has developed a great range of

such illusions, using the simplest equipment—video goggles, mannequins, and rubber arms. By disrupting the usual unity of touch, vision, and proprioception, he has induced uncanny experiences in some people, convincing them that their bodies have shrunk or grown enormous, even that they have swapped bodies with someone else. I experienced this for myself when I visited his laboratory in Stockholm for a number of experiments. In one, I was convinced that I possessed a third arm; in another, I felt embodied in a two-foot-high doll, and as I looked through "its" eyes via video goggles, normal objects in the room appeared enormous.

It is evident, from all of this work, that the brain's representation of the body can often be fooled simply by scrambling the inputs from different senses. If sight and touch say one thing, however absurd, even a lifetime of proprioception and a stable body image cannot always resist this. (Individuals may be more or less susceptible to such illusions, and one might imagine that dancers or athletes, who have an exceptionally vivid sense of where their bodies are in space, may be harder to fool in this way.)

The body illusions Ehrsson is exploring are very much more than party tricks; they point to the ways in which our body ego, our sense of self, is formed from the coordination of senses—not just touch and vision but proprioception and perhaps vestibular sensation, too. Ehrsson and others favor the idea that there are "multisensory" neurons, perhaps at a number of places in the brain, which serve to coordinate the complex (and usually consistent) sensory information coming into the brain. But if this is interfered with—by nature or experiment—our seemingly unassailable certainties about the body and the self can vanish in an instant.

15

# Phantoms, Shadows,
# and Sensory Ghosts

While hallucinations of sight and sound—"visions" and "voices"—are described in the Bible, in *The Iliad* and *The Odyssey*, in all the great epics of the world, none of these so much as mentions the existence of phantom limbs, the hallucinatory feeling that one still has a limb even though it has been amputated. Indeed, there was no term for these before Silas Weir Mitchell gave them their name in the 1870s. And yet they are common—more than a hundred thousand people in the United States have amputations every year, and the vast majority of them experience phantoms after their amputations. The experience of phantom limbs must be as old as amputation itself, and amputations are not new—they were performed thousands of years ago: the Rig Veda tells the story of the warrior queen Vishpla, who went to battle with an iron prosthesis after she lost a leg.

In the sixteenth century, Ambroise Paré, a French military

surgeon who was called upon to amputate dozens of injured limbs, wrote, "Long after the amputation is made, patients say that they still feel pain in the amputated part . . . which seems almost incredible to people who have not experienced this."

Descartes, in his *Meditations on First Philosophy*, observed that, just as the sense of vision was not always reliable, so "errors in judgment" could occur in the "internal senses" as well. "I have sometimes been informed," he wrote, "by parties whose arm or leg had been amputated, that they still occasionally seemed to feel pain in that part of the body which they had lost—a circumstance that led me to think that I could not be quite certain even that any one of my members was affected when I felt pain in it."

But by and large, as the neurologist George Riddoch brought out (in 1941), a curious atmosphere of silence and secrecy seems to surround the subject. "Spontaneous description of phantoms is rarely offered," he wrote. "Dread of the unusual, of disbelief, or even of the accusation of insanity may be behind this reticence."

Weir Mitchell himself hesitated for years before writing professionally on the subject; he introduced it first in the form of fiction (he was a writer as well as a physician), in "The Case of George Dedlow," published anonymously in the *Atlantic Monthly* in 1866. As a neurologist working at a military hospital in Philadelphia during the Civil War (the place was informally known as the "Stump Hospital"), Mitchell saw dozens of amputees and, driven by his own curiosity and compassion, he encouraged them to describe their experiences. It would take him several years to fully digest what he had seen and heard from his patients, but in 1872, in his classic *Injuries of Nerves,* he was able to provide a detailed description and discussion of phantom limbs—the first such in the medical literature.[1]

Mitchell devoted the final chapter of his book to phantom limbs, introducing the subject as follows:

> No history of the physiology of stumps would be complete without some account of the sensorial delusions to which persons are subject in connection with their lost limbs. These hallucinations are so vivid, so strange, and so little dwelt upon by authors, as to be well worthy of study, while some of them seem to me especially valuable, owing to the light which they cast upon the subject of the long-disputed muscular sense.
>
> Nearly every man who loses a limb carries about with him a constant or inconstant phantom of the missing member, a sensory ghost of that much of himself.

---

1. It is likely that there was popular or folk knowledge of the phenomenon long before there were any medical descriptions.

Twenty years before Weir Mitchell named phantom limbs, Herman Melville included a fascinating scene in *Moby-Dick*, where the ship's carpenter is measuring Captain Ahab for a whalebone leg. Ahab addresses the carpenter:

> Look ye, carpenter, I dare say thou callest thyself a right good workmanlike workman, eh? Well, then, will it speak thoroughly well for thy work, if, when I come to mount this leg thou makest, I shall nevertheless feel another leg in the same identical place with it; that is, carpenter, my old lost leg; the flesh and blood one, I mean. Canst thou not drive that old Adam away?
>
> [The carpenter replies:] Truly, sir, I begin to understand somewhat now. Yes, I have heard something curious on that score, sir; how that a dismasted man never entirely loses the feeling of his old spar, but it will be still pricking him at times. May I humbly ask if it be really so, sir?
>
> It is, man [says Ahab]. Look, put thy live leg here in place where mine once was; so, now, here is only one distinct leg to the eye, yet two to the soul. Where thou feelest tingling life; there, exactly there, there to a hair, do I.

After Mitchell had brought attention to the subject, other neurologists and psychologists were drawn to study phantom limbs. Among them was William James, who sent a questionnaire to eight hundred amputees (he was able to contact them with the help of prosthetic manufacturers), and of these, nearly two hundred answered the questionnaire; a few he was able to interview personally.[2]

Where Mitchell's observations, working with Civil War amputees, were of fresh, just developed phantom limbs, James was able to study a much more varied population (one man, in his seventies, had had a thigh amputation sixty years earlier), and so he was in a better position to describe the changes in phantom limbs over years or decades, changes which he described in detail in his 1887 paper on "The Consciousness of Lost Limbs."

James was especially interested in the way that initially vivid and mobile phantoms often tended to shorten or disappear with time. This surprised him more than the presence of phantoms, which he felt was only to be expected with continuing activity in the areas of the brain that represented sensation and movement in the lost limb. "The popular mind wonders how the lost feet can still be felt," James wrote. "For me, the cause for wonder are those in which the lost feet are not felt." Hand phantoms, he observed, unlike leg or arm phan-

---

2. The importance of first-person accounts was emphasized by William James in his 1887 paper "The Consciousness of Lost Limbs":

> In a delicate inquiry like this, little is to be gained by distributing circulars. A single patient with the right sort of lesion and a scientific mind, carefully cross-examined, is more likely to deepen our knowledge than a thousand circulars answered as the average patient answers them, even though the answers be never so thoroughly collated by the investigator.

toms, rarely disappeared. (We now know that this is because the fingers and hands have a particularly massive representation in the brain.) He did, however, note that the intervening arm might disappear, so that the preserved phantom hand now seemed to sprout from the shoulder.[3]

He was also struck by the way in which an initially mobile phantom could become immobile or even paralyzed, so that "no effort of will can make it change [its position]." (In rare cases, he said, "the very attempt to will the change has grown impossible.") James saw that fundamental questions were raised here about the neurophysiology of "will" and "effort," though he could not answer them. And they were not to be answered for more than a century, until V. S. Ramachandran clarified the nature of "learned" paralysis in phantom limbs in the 1990s.

P hantom limbs are hallucinations insofar as they are perceptions of something that has no existence in the outside world, but they are not quite comparable to hallucinations of sight and sound. While losing one's eyesight or hearing may lead to corresponding hallucinations in 10 or 20 percent of those affected, phantom limbs occur in virtually all who have had a limb amputated. And while it may be months or years before hallucinations follow blindness or deafness, phantom limbs appear immediately or within days after an amputation—and

---

3. The reason for this was not to be clarified until a century later, when it became possible to visualize, with fMRI, the gross changes in the brain's body mapping that could occur after an amputation. Michael Merzenich and his colleagues at UCSF, working with both monkeys and humans, have shown how rapid and radical such changes may be.

they are felt as an integral part of one's own body, unlike any other sort of hallucination. Finally, while visual hallucinations such as those of Charles Bonnet syndrome are varied and full of invention, a phantom closely resembles the physical limb that was amputated in size and shape. A phantom foot may have a bunion, if the real one did; a phantom arm may wear a wristwatch, if the real arm did. In this sense, a phantom is more like a memory than an invention.

The near universality of phantom limbs after amputation, the immediacy of their appearance, and their identity with the corporeal limbs in whose stead they appear suggest that, in some sense, they are already in place—revealed, so to speak, by the act of amputation. Complex visual hallucinations get their material from the visual experiences of a lifetime—one has to have seen people, faces, animals, landscapes to hallucinate them; one has to have heard pieces of music to hallucinate them. But the feeling of a limb as a sensory and motor part of oneself seems to be innate, built-in, hardwired—and this supposition is supported by the fact that people born without limbs may nonetheless have vivid phantoms in their place.[4]

---

4. Despite categorical assertions by many that "congenital" phantoms cannot occur, there have been several reports (as Scatena has noted in a review of the subject) indicating that some people with aplasia—congenitally defective or absent limbs—do have phantoms. Klaus Poeck, in 1964, described an eleven-year-old girl born without forearms or hands who was able to "move" her phantom hands. As Poeck wrote, "In her first years at school, she had learned to solve simple arithmetic problems by counting with her fingers. . . . On these occasions she would place her phantom hands on the table and count the outstretched fingers one by one."

It is not clear why some people born without limbs have phantoms and some do not. What is clear, as Funk, Shiffrar, and Brugger observed in one study, is that those who do have phantoms seem to have cerebral "action observation systems" similar to those of normally limbed people, allowing them to grasp action patterns by observing others and

The most fundamental difference between phantom limbs and other hallucinations is that they can be moved voluntarily, whereas visual and auditory hallucinations proceed autonomously, outside one's control. This was also emphasized by Weir Mitchell:

> [The majority of amputees] are able to will a movement, and apparently they themselves execute it more or less effectively. . . . The certainty with which these patients describe their [phantom motions], and their confidence as to the place assumed by the parts moved, are truly remarkable . . . the effect is apt to excite twitching in the stump. . . . In some cases the muscles which act on the hand are absent altogether; yet in these cases there is fully as clear and definite a consciousness of the movement of the fingers and of their change of positions as in cases [where the muscles of the hand are partially preserved].

Other hallucinations are only sensations or perceptions, albeit of a very special sort, whereas a phantom limb is capable of phantom *action*. Given a suitable prosthesis, the phantom limb will slip into the prosthesis ("like a hand into a glove," as many patients say)—slip into it and animate it, so that the artificial limb can be used like a real one. Indeed, this must happen if one is to use a prosthesis effectively. The artificial limb becomes part of one's body, of one's body image, as a cane in a blind man's hand becomes an extension of himself. One may say that an artificial leg, for instance, "clothes" the phantom, allows it to be effective, gives it an objective sensory and motor existence, so that it can often "feel" and respond to

---

to internalize these as mobile phantoms. Those born without limbs who do not have phantoms, Funk et al. propose, may have problems in motion perception, especially judging the movements of other people's limbs.

minute irregularities in the ground almost as well as the original leg.[5] (Thus the great climber Geoffrey Winthrop Young, who lost a leg during World War I, was able to climb the Matterhorn using a prosthetic limb of his own design.)[6]

One might go further and say that a phantom is a portion of body image which is lost or dissociated from its natural, embodying home (the body)—and, as such, as something extraneous, it may be intrusive or deceptive (thus the danger of walking off a curb with a phantom leg). The lost phantom (if one can speak figuratively) longs for a new home, and it will find this in a suitable prosthesis. I have had many patients tell me how they may be disturbed by their phantom at night but relieved in the morning, for the phantom disappears the moment they put on their prosthesis—disappears, that is, *into* the prosthesis, merging so seamlessly with it that phantom and prosthesis become one.

Knowledge of what one is doing with one's phantom—even without a prosthesis—can be exquisitely refined. As a young student, Erna Otten, a distinguished pianist, was a pupil of the great Paul Wittgenstein, who lost his right arm in the First

---

5. When Henry Head introduced the term "body image" (fifty or so years after Weir Mitchell had introduced the term "phantom limb"), he did not mean it to refer to a purely sensory image or map in the brain—he had in mind an image or model of agency and action, and it is this which needs to be embodied in an artificial limb.

Philosophers like to speak of "embodiment" and "embodied agency," and there is no simpler place to study this than in the nature of phantoms and their embodiment in artificial limbs—prosthesis and phantom go together like body and soul. I have wondered whether some of Ludwig Wittgenstein's philosophical notions were suggested by his brother's phantom arm—thus his final work, *On Certainty*, starts from the certainty of the body, the body as embodied agency.

6. Wade Davis describes this in his book *Into the Silence: The Great War, Mallory, and the Conquest of Everest.*

World War but continued to play with his left hand (and com-
missioned a number of composers to write music for the left
hand). Yet he continued to teach, in a sense, with both hands.
In a letter to the *New York Review of Books,* responding to an
article I had written, Otten wrote:

> I had many occasions to see how involved his right stump was
> whenever we went over the fingering for a new composition. He
> told me many times that I should trust his choice of fingering
> because he felt every finger of his right hand. At times I had to
> sit very quietly while he would close his eyes and his stump
> would move constantly in an agitated manner. This was many
> years after the loss of his arm.

Unfortunately, not all phantoms are as well formed, as pain-
less, or as mobile as Wittgenstein's. Many show a tendency to
shrink or "telescope" with time—a phantom arm may be reduced
to a hand seemingly sprouting from the shoulder. This tendency
to shrink is minimized by embedding the phantom in a prosthesis and using it as much as possible. A phantom may also become
paralyzed or contorted in painful positions, with its "muscles"
in spasm. Thus Admiral Lord Nelson, after losing his right arm
in battle, developed a phantom limb with the hand permanently
clenched, the fingers digging excruciatingly into the palm.[7]

---

7. Nonetheless, Nelson regarded his phantom as "a direct proof for the
existence of the soul." The survival of a spiritual arm after a corporeal
one was annihilated, he thought, epitomized the survival of the soul
after bodily death.
    For Captain Ahab, however, this was a matter for horror as much as
wonder: "And if I still feel the smart of my crushed leg, though it be now
so long dissolved; then, why mayst not thou, carpenter, feel the fiery
pains of hell forever, and without a body? Hah!"

Such disorders of body image have long seemed inexplicable and untreatable. But over the last few decades, it has become clear that the body image is not as fixed as we once thought; indeed, it is remarkably plastic, and extensive reorganization or remapping can occur with phantom limbs.

If there is interruption of nerve function from injury or disease in the spinal cord or peripheral nerves, cutting off or reducing normal sensory input to the brain, this may cause major disturbance of body image, with strange phantom images superimposed on the real but insentient body parts. This was very striking with a colleague of mine, Jeannette W., who broke her neck in a car accident and became quadriplegic, with a complete absence of sensation below the level of the fracture. She had, in a sense, been "amputated" from the neck down and had little sense of her body below this. But in its place, she had a phantom body, which was unstable and prone to distortions and deformations. She could reverse these, for a while, by *seeing* that her body still had a normal shape and conformation, and she arranged for mirrors to be set up in her office and in the hospital corridors, so that she could glance up and (in her words) take "visual sips" from them as she bowled past in her wheelchair.

As normal sensation is blocked, body image disturbances can occur very quickly. Most of us have had strange phantom experiences with dental anesthesia, of a grotesquely swollen, deformed, or misplaced cheek or tongue. Looking in a mirror will do little to dispel these illusions, which disappear only with the return of normal sensation. One patient of mine, with the removal of a large brain tumor, had to sacrifice the roots of the sensory nerves on one side of her face. For years following this, she had a persistent sense that the whole right side of her face was "slipping," "caved in," or "missing"; that her tongue and cheek on this side were tremendously swollen and grotesque-looking. She later came to have a leg amputated, and soon after surgery became aware of a phantom leg. Now, she said, "I know what's wrong with my face. It's exactly the same feeling—I have a phantom face."

There can also be extra limbs—supernumerary phantoms— if certain areas of the body are denervated. A striking example of this was described by Richard Mayeux and Frank Benson. Their patient was a young man with multiple sclerosis who developed a numbness on his right side and then experienced, as they wrote,

> a tactile illusion that a second right arm was lying across his lower chest and upper abdomen. The extra arm seemed to be attached to the chest wall . . . . There was only a vague sensation of the duplicate illusory lower forearm, wrist, and palm, but a vivid impression of the fingers lying on the abdominal wall. . . . The illusion persisted for period of 5 to 30 minutes and was accompanied by a "gripping" sensation of the illusory hand. . . . The phantom limb sensation was always coincident with feelings of increased stiffness, numbness, and burning [sensations] of the actual right arm.

Nelson's clenched hand exemplifies an unpleasant evolution which phantom limbs may undergo—phantoms which are initially loose, mobile, and obedient to the will may subsequently become paralyzed, contorted, and often intensely painful. Before the 1990s, there was no plausible explanation as to why phantom limbs might get frozen in this way, nor any notion of how to unfreeze them. But in 1993, V. S. Ramachandran suggested a physiological scenario which might explain the progressive loss of voluntary movement so common in phantom limbs. The vivid sense that one could move a phantom limb freely, he thought, went with the brain being able to monitor its own motor commands to the phantom. But with the continuing absence of visual or proprioceptive confirmation of movement, the brain, in effect, might "abandon" the limb. Thus, Ramachandran thought, paralysis was "learned," and he wondered whether it could be unlearned.

Could one, by simulating visual and proprioceptive feedback, dupe the brain into believing that the phantom was once again mobile and capable of voluntary movement? Ramachandran developed a brilliantly simple device—an oblong wooden box with its left and right sides divided by a mirror, so that looking into the box from one side or the other, one would get an illusion of seeing both hands, where in reality one was seeing only one hand and its mirror image. Ramachandran tried this device on a young man who had had a partial amputation of his left arm—his now-rigid phantom hand, Ramachandran wrote, "jutted like a mannequin's resin-case forearm out of the stump. Far worse, it was also subject to painful cramping that his doctors could do nothing about."

After explaining what he had in mind, Ramachandran asked the young man to "insert" his phantom arm to the left of the mirror. Ramachandran described this in his book *The Tell-Tale Brain:*

> He held out his paralyzed phantom on the left side of the mirror, looked into the right side of the box and carefully positioned his

right hand so that its image was congruent with (superimposed on) the felt position of the phantom. This immediately gave him the startling visual impression that the phantom had been resurrected. I then asked him to perform mirror-symmetric movements of both arms and hands while he continued looking into the mirror. He cried out, "It's like it's plugged back in!" Now he not only had a vivid impression that the phantom was obeying his commands, but to his amazement, it began to relieve his painful phantom spasms for the first time in years. It was as though the mirror visual feedback (MVF) had allowed his brain to "unlearn" the learned paralysis.

This extremely simple procedure (which was devised only after much careful thinking and a whole, very original theory as to the many interacting factors involved in the production of phantoms and their vicissitudes) can easily be modified for dealing with phantom legs and a variety of other conditions involving distortion of body image.

The *appearance* of the hand moving, the optical illusion, was sufficient to generate the *feeling* that it was moving. I described the converse of this in *The Mind's Eye*, when the existence of a large blind spot in my visual field allowed me, visually, to "amputate" a hand. But if, when I had done this, I opened and closed my fist or moved my now-invisible fingers, a sort of pink protoplasmic extension grew out of my visual "stump" and developed into a (visual) phantom of the hand.

Jonathan Cole and his colleagues have made similar observations, testing a virtual reality system to reduce phantom pain. In their experiments with leg and arm amputees, the amputated stump is connected to a motion capture device, which in turn determines the movements of a virtual arm or

leg on a computer screen. Most of their subjects learned to correlate their own movements with those of the on-screen ava-
tar, and developed a sense of agency or ownership, so that they
were able to move the virtual limb with surprising delicacy
(for instance, to reach for and grasp a virtual apple lying on the
surface of a virtual table). Such learning occurred remarkably
quickly, within half an hour or so. With this sense of agency
and intentionality often came a reduction in phantom pain—
and even virtual perception. One man, for example, could
"feel" the virtual apple when he picked it up. Cole and his colleagues wrote, "Perception was not only of motion of the limb
but also of touch, a virtual-visual cross-modal perception."

In 1864, Weir Mitchell and two of his colleagues put out a
special circular from the Surgeon General's Office, entitled
*Reflex Paralysis*. In reflex paralysis, the injured limb is intact,
but it cannot be moved; it seems absent or "alien," not part of
the body. It is, in a sense, the opposite of a phantom limb—an
external limb with no internal image to give it presence and
life.

I had such an experience in 1974 during the mountaineer-
ing accident in which I ruptured the quadriceps tendon in my
left leg. Though the tendon was repaired surgically, there was
damage at the neuromuscular junction, and additionally, the
leg was hidden from sight and touch, immobilized in a long,
opaque cast. Under these circumstances, where it was impossible to send commands to the injured muscle and there was
no sensory or visual feedback, the leg disappeared from my
body image, leaving (so it seemed to me) an inanimate, alien
thing in its place. This continued to be the case for thirteen

days. (Thinking back on this experience, I wonder whether one of Ramachandran's mirror boxes would have helped me to recover movement, and a sense of reality, in this leg sooner. It might have helped, too, had the cast been transparent, so that I could at least see the leg.)

It was an experience so uncanny that I wrote an entire book, *A Leg to Stand On*, about it. I suggested, only half-jokingly, that readers would more easily imagine such experiences if they read the book under spinal anesthesia, for as the anesthetic blocks activity in the spinal cord, one's lower half becomes not only paralyzed and senseless but, subjectively, nonexistent. One feels that one's body terminates in the middle, and that what lies below—hips and a pair of legs—do not belong to one; they could just as well be a wax model from an anatomy museum. This lack of ownership, this alienation, is bizarre to experience. I found it almost intolerable during the thirteen days in which my left leg seemed alien to me—I wondered, darkly, whether any recovery would occur and whether, if it did not, I would do best to have the useless leg removed.

There may indeed, though very rarely, be a congenital absence of body image in an otherwise normal limb; this is suggested, at least, by the numerous reported cases of what Peter Brugger has termed "body-integrity identity disorder." Such people feel, from childhood onward, that one of their limbs, or perhaps a part of a limb, is not theirs, but an alien encumbrance, and this feeling may engender a passionate desire to have the "superfluous" limb amputated.

Prior to 1990, the whole field of phantom limbs and other disturbances of body image could be studied only phenomenologically, from the accounts and behaviors of those afflicted. Such conditions were often ascribed to hysteria or an overactive imagination, but the development of sophisticated brain imaging has changed this by showing the physiological changes in the brain (especially in parts of the parietal lobes) which underlie such strange experiences. This, along with ingenious experiments such as Ramachandran's mirror box, has allowed us to get a clearer view of the neural basis of embodiment, of agency, of self; to bring purely clinical and sometimes purely philosophical ideas into the realm of neuroscience.

S hadows" and "doubles"—hallucinatory distortions of the body and body image—take us into an even stranger realm. If a limb or part of the body is "deanimated" by nerve or spinal cord damage, the deanimated part itself may feel lifeless, inorganic, alien. But if there is damage to the right parietal lobe, a much deeper form of estrangement may occur. The deanimated part of the body—if its existence is acknowledged at all—is felt to belong to someone else, a mysterious "other." Many years ago, as a medical student, I saw a patient who had been admitted to the neurosurgery service for removal of a parietal lobe tumor. One evening, while awaiting surgery, he fell out of bed in a peculiar way—almost, the nurses said, as if he had thrown himself off the bed. When I asked him about this, he said that he had been asleep and awoke to discover a leg—a dead, cold, hairy leg—in his bed. He could not think how someone else's leg had got into his bed, unless—the idea suddenly occurred to him—the nurses had taken a leg from the anatomy labs and slipped it into his bed as a joke. Shocked and repelled, he used his good right leg to kick the alien thing out of his bed, and, of course, he came out after it, and was now aghast because "it" was attached to him. I said, "But it is *your* leg," and pointed out to him that the size, the shape, the contour, the color were precisely the same in the two legs; but he would have none of it. He was absolutely certain that it was someone else's.[8]

Over the years I have seen other patients who, in consequence of a right-hemisphere stroke, have lost all feeling and use of the left side. Often they have no awareness that anything has happened, but some people are convinced that their left side belongs to someone else ("my twin brother," "the man next to me," even "It's *yours*, Doc, who are you kidding?"). Perhaps "my twin brother" is a hieroglyphic way of indicating that while half of the body seems alien, it also seems very akin, almost identical to oneself . . . that it *is* oneself in a strange, disguised way. It needs to be emphasized that such patients may be highly intelligent, lucid, and articulate—and that it is solely in reference to their odd distortions of body image that they make their surreal but irrefragable statements.

The feeling that *someone is there,* to the left or the right, perhaps just behind us, is known to us all. It is not just a vague feeling; it is a distinct sensation. We may wheel around to catch the lurking figure, but there is no one to be seen. And yet it is impossible to dismiss the sensation, even if we have learned from repeated experience that this sort of sensed presence is a hallucination or an illusion.

The sensation is commoner if one is alone, in darkness, perhaps in unfamiliar surroundings, hyperalert. It is well known to mountaineers and polar explorers, where the vastness and

---

8. This story, "The Man Who Fell Out of Bed," is related more fully in *The Man Who Mistook His Wife for a Hat.*

danger of the terrain, the isolation and exhaustion (and, in the mountains, reduced oxygen) contribute to the feeling. The sensed presence, the invisible companion, the "third man," the shadow person—all sorts of terms are used—is well aware of us, and has definite intentions, whether these are benign or malignant. The shadow stalking us has something in mind. And it is this sense of its intentionality or agency which either raises the hair on our neck or produces a sweet, calm feeling of being protected, not alone.

While the sense of "somebody there" is commoner in the hypervigilant states induced by some forms of anxiety, by various drugs and by schizophrenia, it may also occur in neurological conditions. Thus Professor R. and Ed W., who both have advancing Parkinson's disease, have persistent feelings of a presence—something or someone they never actually see; this presence is always on the same side. There may be a transitory sense of "someone there" in attacks of migraine or in seizures—but a very persistent sense of a presence, always to the same side, is suggestive of a brain lesion. (This is also the case with such experiences as déjà vu, which we all have occasionally, but which, if very frequent, suggests a seizure disorder or a brain lesion.)

In 2006 Olaf Blanke and his colleagues (Shahar Arzy et al.) described how, with a young woman being evaluated for surgical treatment of epilepsy, they could predictably induce a "shadow-person" by electrical stimulation of the left temporoparietal junction. When the woman was lying down, a mild stimulation of this area gave her the impression that someone was behind her; a stronger stimulation allowed her to define the "someone" as young but of indeterminate sex, lying down in a position identical to her own. When stimulations were repeated with her in a sitting position, embracing her knees with her arms, she sensed a man behind her, sitting in the same position and clasping her with his intangible arms. When she was given a card to read for a language learning test, the sitting "man" moved to her right side, and she understood that he had aggressive intentions ("He wants to take the card. . . . He doesn't want me to read."). There were thus elements of the "self" here—the mimicking or sharing of her postures by the shadow person—as well as elements of the "other."[9]

That there may be some connection between body-image disturbances and hallucinatory "presences" was brought out as early as 1930 by Engerth and Hoff, as Blanke and his colleagues wrote in a 2006 paper. Engerth and Hoff described an elderly man who had become hemianopic after a stroke. He saw "silver things" in the blind half of his visual field, then automobiles coming at him from the left, and then people: "countless" people, all identical in appearance and with a clumsy gait, staggering, with the right arm outstretched—precisely the gait the patient himself had when he tried to walk and avoid colliding with people on his left.

But he also had alienation of his left side, and he felt that this side of his body was "filled with something strange."

"Finally," Engerth and Hoff wrote, "the host of hallucinations disappeared, and there then appeared what the patient called 'a constant companion.' Wherever the patient went, he saw someone walking along on his left. . . . At the moment

---

9. Several people have written to me with similar stories of sensing a presence just as they are going to sleep or waking. Linda P. observed that once, as she was drifting off to sleep, she felt "as if I was being held on my right side, as if someone had put their arms around me and was stroking my hair. It was a lovely feeling; then I remembered that I was alone, and [the feeling disappeared]."

when the companion appeared, the alien feeling in the left half of the body disappeared. . . . We would not be in error," they concluded, "if we saw in this 'companion' the left half of the body which had become independent."

It is not clear whether this "constant companion" is to be classified as a "sensed presence" or an autoscopic "double"—it has qualities of both. And perhaps some of these seemingly distinct categories of hallucination merge. Blanke and his colleagues, writing in 2003 of body-image, or "somatognosic," disorders, observed that these may take a number of forms: illusions of a missing body part, a transformed (enlarged or shrunk) body part, a dislocated or disconnected body part, a phantom limb, a supernumerary limb, an autoscopic image of one's own body, or a "feeling of a presence." All of these disorders, Blanke stresses, with their hallucinations of vision, touch, and proprioception, are associated with parietal or temporal lobe damage.

J. Allan Cheyne has also investigated sensed presences, both in the relatively mild form that may occur when one is fully conscious and in the terrifying form that is often associated with sleep paralysis. He speculates that this feeling of "presence"—a universal human (and perhaps animal) sensation—may have a biological origin in "the activation of a distinct and evolutionary functional 'sense of the other' . . . deep within the temporal lobe specialized for the detection of cues for agency, especially those potentially associated with threat or safety."

Sensed presence not only has its place in the neurological literature; it also forms a chapter in William James's *Varieties of Religious Experience*. He recounts a number of case histories where the initially horrible feeling of an intrusive and threatening "presence" became a joyful and even blissful one, including that of a friend who told him:

> It was about September, 1884, when I had the first experience . . . suddenly I FELT something come into the room and stay close to my bed. It remained only a minute or two. I did not recognize it by any ordinary sense and yet there was a horribly unpleasant "sensation" connected with it. It stirred something more at the roots of my being than any ordinary perception. . . . Something was present with me, and I knew its presence far more surely than I have ever known the presence of any fleshly living creature. I was as conscious of its departure as of its coming: an almost instantaneously swift going through the door, and the "horrible sensation" disappeared. . . .
>
> [On a subsequent occasion], there was not a mere consciousness of something there, but fused in the central happiness of it, a startling awareness of some ineffable good. Not vague either, not like the emotional effect of some poem, or scene, or blossom, or music, but the sure knowledge of the close presence of a sort of mighty person.

"Of course," added James, "such an experience as this does not connect itself with the religious sphere . . . [and] my friend . . . does not interpret these latter experiences theistically, as signifying the presence of God."

But one can readily see why others, perhaps of a different disposition, might interpret the "sure knowledge of the close presence of a sort of mighty person" and "a startling awareness of some ineffable good" in mystical, if not religious, terms. Other case histories in James's chapter bear this out, leading him to say that "many persons (how many we cannot tell) possess the objects of their belief not in the form of mere conceptions which the intellect accepts as true, but rather in the form of quasi-sensible realities directly apprehended."

Thus the primal, animal sense of "the other," which may have evolved for the detection of threat, can take on a lofty, even transcendent function in human beings, as a biological basis for religious passion and conviction, where the "other," the "presence," becomes the person of God.

# Acknowledgments

I am most grateful, first and foremost, to the hundreds of patients and correspondents who have shared their experiences of hallucinations with me over many decades, and especially to those who have allowed me to quote their words and tell their stories in this book.

I owe an enormous debt to my friend and colleague Orrin Devinsky, who has stimulated my thoughts with his many published and forthcoming papers and referred many of his patients to me. I have enjoyed and benefited from discussions with Jan Dirk Blom and from reading his wonderfully comprehensive *Dictionary of Hallucinations* and *Hallucinations: Research and Practice.* I am deeply grateful for the friendship and advice of my colleagues Sue Barry, Bill Borden, William Burke, Kevin Cahill, Jonathan Cole, Douwe Draaisma, Henrik Ehrsson, Dominic ffytche, Steven Frucht, Mark Green, James Lance, Richard Mayeux, Alvaro Pascual-Leone, Stanley Prusiner, V. S. Ramachandran, and Leonard Shengold. And I am grateful to Gale Delaney, Andreas Mavromatis, Lylas Mogk, Jeff Odel, and Robert Teunisse for sharing their own experiences (and sometimes patients) with me.

I must also thank Molly Birnbaum, Daniel Breslaw, Leslie Burkhardt, Elizabeth Chase, Allen Furbeck, Kai Furbeck, Ben Helfgott, Richard Howard, Hazel Rossotti, Peter Selgin, Amy Tan, Bonnie Thompson, Kappa Waugh, and Edward Weinberger. Eveline Honig, Audrey Kindred, Sharon Smith, and others at the Narcolepsy Network kindly introduced me to many people with narcolepsy and sleep paralysis. Bill Hayes, a friend and a writer whom I much admire, read each chapter with his own writerly eye and made many valuable suggestions.

For their support and encouragement, I thank David and Susie Sainsbury; Dan Frank, who has patiently reviewed draft after draft of this book (as with many previous ones); Hailey Wojcik, invaluable research assistant, typist, and swimming companion; and Kate Edgar, my friend, editor, and collaborator for thirty years, to whom this book is dedicated.

# Bibliography

Abell, Truman. 1845. Remarkable case of illusive vision. *Boston Medical and Surgical Journal* 33 (21): 409–13.

Adair, Virginia Hamilton. 1996. *Ants on the Melon: A Collection of Poems*. New York: Random House.

Adamis, Dimitrios, Adrian Treloar, Finbarr C. Martin, and Alastair J. D. Macdonald. 2007. A brief review of the history of delirium as a mental disorder. *History of Psychiatry* 18 (4): 459–69.

Adler, Shelley R. 2011. *Sleep Paralysis: Night-mares, Nocebos, and the Mind-Body Connection*. Piscataway, NJ: Rutgers University Press.

Airy, Hubert. 1870. On a distinct form of transient hemiopsia. Communicated by the Astronomer Royal. *Philosophical Transactions of the Royal Society of London* 160: 247–64.

Alajouanine, T. 1963. Dostoiewski's epilepsy. *Brain* 86 (2): 209–18.

Ardis, J. Amor, and Peter McKellar. 1956. Hypnagogic imagery and mescaline. *British Journal of Psychiatry* 102: 22–29.

Arzy, Shahar, Gregor Thut, Christine Mohr, Christoph M. Michel, and Olaf Blanke. 2006. Neural basis of embodiment: Distinct contributions of temporoparietal junction and extrastriate body area. *Journal of Neuroscience* 26 (31): 8074–81.

Asheim, Hansen B., and Eylert Brodtkorb. 2003. Partial epilepsy with "ecstatic" seizures. *Epilepsy & Behavior* 4 (6): 667–73.

Baethge, Christopher. 2002. Grief hallucinations: True or pseudo? Serious or not? An inquiry into psychopathological and clinical features of a common phenomenon. *Psychopathology* 35: 296–302.

Bartlett, Frederic C. 1932. *Remembering: A Study in Experimental and Social Psychology*. Cambridge: Cambridge University Press.

Baudelaire, Charles. 1860/1995. *Artificial Paradises*. New York: Citadel.

Berrios, German E. 1981. Delirium and confusion in the nineteenth century: A conceptual history. *British Journal of Psychiatry* 139: 439–49.

Bexton, William H., Woodburn Heron, and T. H. Scott. 1954. Effects of decreased variation in the sensory environment. *Canadian Journal of Psychology* 8 (2): 70–76.

Birnbaum, Molly. 2011. *Season to Taste: How I Lost My Sense of Smell and Found My Way.* New York: Ecco / Harper Collins.

Blanke, Olaf, Stéphanie Ortigue, Alessandra Coeytaux, Marie-Dominique Martory, and Theodor Landis. 2003. Hearing of a presence. *Neurocase* 9 (4): 329–39.

Blanke, Olaf, Shahar Arzy, Margitta Seeck, Stephanie Ortigue, and Laurent Spinelli. 2006. Induction of an illusory shadow person. *Nature* 443: 287.

Bleuler, Eugen. 1911/1950. *Dementia Praecox; or, The Group of Schizophrenias.* Oxford: International Universities Press.

Blodgett, Bonnie. 2010. *Remembering Smell: A Memoir of Losing—and Discovering—the Primal Sense.* New York: Houghton Mifflin Harcourt.

Blom, Jan Dirk. 2010. *A Dictionary of Hallucinations.* New York: Springer.

Blom, Jan Dirk, and Iris E. C. Sommer, eds. 2012. *Hallucinations: Research and Practice.* New York: Springer.

Bonnet, Charles. 1760. *Essai analytique sur les facultés de l'âme.* Copenhagen: Freres Cl. & Ant. Philibert.

Boroojerdi, Babak, Khalaf O. Bushara, Brian Corwell, Ilka Immisch, Fortunato Battaglia, Wolf Muellbacher, and Leonardo G. Cohen. 2000. Enhanced excitability of the human visual cortex induced by short-term light deprivation. *Cerebral Cortex* 10: 529–34.

Botvinick, Matthew, and Jonathan Cohen. 1998. Rubber hands "feel" touch that eyes see. *Nature* 391: 756.

Brady, John Paul, and Eugene E. Levitt. 1966. Hypnotically induced visual hallucinations. *Psychosomatic Medicine* 28 (4): 351–63.

Brann, Eva. 1993. *The World of the Imagination: Sum and Substance.* Lanham, MD: Rowman & Littlefield.

Brewin, Chris, and Steph J. Hellawell. 2004. A comparison of flashbacks and ordinary autobiographical memories of trauma: Content and language. *Behaviour Research and Therapy* 42 (1): 1–12.

Brierre de Boismont, A. 1845. *Hallucinations; or, The Rational History of Apparitions, Visions, Dreams, Ecstasy, Magnetism and Somnambulism.* First English edition, 1853. Philadelphia: Lindsay and Blakiston.

Brock, Samuel. 1928. Idiopathic narcolepsy, cataplexia and catalepsy associated with an unusual hallucination: A case report. *Journal of Nervous and Mental Disease* 68 (6): 583–90.

Brugger, Peter. 2012. Phantom limb, phantom body, phantom self. A phenomenology of "body hallucinations." In *Hallucinations: Research and Practice*, ed. Jan Dirk Blom and Iris E. C. Sommer. New York: Springer.

Brugger, Peter, R. Agosti, M. Regard, H. G. Wieser, and T. Landis. 1994. Heautoscopy, epilepsy, and suicide. *Journal of Neurology, Neurosurgery and Psychiatry* 57: 838–39.

Burke, William. 2002. The neural basis of Charles Bonnet hallucinations: A hypothesis. *Journal of Neurology, Neurosurgery and Psychiatry* 73: 535–41.

Carlson, Laurie Winn. 1999. *A Fever in Salem: A New Interpretation of the New England Witch Trials*. Chicago: Ivan R. Dee.

Cheyne, J. Allan. 2001. The ominous numinous: Sensed presence and "other" hallucinations. *Journal of Consciousness Studies* 8 (5–7): 133–50.

———. 2003. Sleep paralysis and the structure of waking-nightmare hallucinations. *Dreaming* 13 (3): 163–79.

Cheyne, J. Allan, Steve D. Rueffer, and Ian R. Newby-Clark. 1999. Hypnagogic and hypnopompic hallucinations during sleep paralysis: Neurological and cultural construction of the night-mare. *Consciousness and Cognition* 8 (3): 319–37.

Chodoff, Paul. 1963. Late effects of the concentration camp syndrome. *Archives of General Psychiatry* 8 (4): 323–33.

Cogan, David G. 1973. Visual hallucinations as release phenomena. *Albrecht von Graefes Archiv für klinische und experimentelle Ophthalmologie* 188 (2): 139–50.

Cole, Jonathan, Oliver Sacks, and Ian Waterman. 2000. On the immunity principle: A view from a robot. *Trends in Cognitive Sciences* 4 (5): 167.

Cole, Jonathan, Simon Crowle, Greg Austwick, and David Henderson Slater. 2009. Exploratory findings with virtual reality for phantom limb pain; from stump motion to agency and analgesia. *Disability and Rehabilitation* 31 (10): 846–54.

Cole, Monroe. 1999. When the left brain is not right the right brain may be left: Report of personal experience of occipital hemianopia. *Journal of Neurology, Neurosurgery and Psychiatry* 67: 169–73.

Critchley, Macdonald. 1939. Neurological aspect of visual and auditory hallucinations. *British Medical Journal* 2 (4107): 634–39.

———. 1951. Types of visual perseveration: "Paliopsia" and "illusory visual spread." *Brain* 74: 267–98.

———. 1967. Migraine: From Cappadocia to Queen Square. In *Background to Migraine*, ed. Robert Smith. London: William Heinemann.

Daly, David. 1958. Uncinate fits. *Neurology* 8: 250–60.

Davies, Owen. 2003. The nightmare experience, sleep paralysis, and witchcraft accusations. *Folklore* 114 (2): 181–203.

Davis, Wade. 2011. *Into the Silence: The Great War, Mallory, and the Conquest of Everest*. New York: Knopf.

de Morsier, G. 1967. Le syndrome de Charles Bonnet: Hallucinations visuelles des vieillards sans déficience mentale. *Annales Médico-Psychologiques* 125: 677–701.

Dening, T. R., and German E. Berrios. 1994. Autoscopic phenomena. *British Journal of Psychiatry* 165: 808–17.

De Quincey, Thomas. 1822. *Confessions of an English Opium-Eater*. London: Taylor and Hessey.

Descartes, René. 1641/1960. *Meditations on First Philosophy*. New York: Prentice Hall.

Devinsky, Orrin. 2009. Norman Geschwind: Influence on his career and comments on his course on the neurology of behavior. *Epilepsy & Behavior* 15 (4): 413–16.

Devinsky, Orrin, and George Lai. 2008. Spirituality and religion in epilepsy. *Epilepsy & Behavior* 12 (4): 636–43.

Devinsky, Orrin, Edward Feldman, Kelly Burrowes, and Edward Bromfield. 1989. Autoscopic phenomena with seizures. *Archives of Neurology* 46 (10): 1080–88.

Devinsky, O., L. Davachi, C. Santchi, B. T. Quinn, B. P. Staresina, and T. Thesen. 2010. Hyperfamiliarity for faces. *Neurology* 74 (12): 970–74.

Dewhurst, Kenneth, and A. W. Beard. 1970. Sudden religious conversions in temporal lobe epilepsy. *British Journal of Psychiatry* 117: 497–507.

Dewhurst, Kenneth, and John Pearson. 1955. Visual hallucinations of the self in organic disease. *Journal of Neurology, Neurosurgery, and Psychiatry* 18: 53–57.

Dickens, Charles. 1861. *Great Expectations*. London: Chapman and Hall.

Dostoevsky, Fyodor M. 1869/2002. *The Idiot.* New York: Everyman's Library

———. 1846/2005. *The Double* and *The Gambler.* New York: Everyman's Library.

Draaisma, Douwe. 2009. *Disturbances of the Mind.* New York: Cambridge University Press.

Ebin, David, ed. 1961. *The Drug Experience: First-Person Accounts of Addicts, Writers, Scientists and Others.* New York: Orion.

Efron, Robert. 1956. The effect of olfactory stimuli in arresting uncinate fits. *Brain* 79 (2): 267–81.

Ehrsson, H. Henrik. 2007. The experimental induction of out-of-body experiences. *Science* 317 (5841): 1048.

Ehrsson, H. Henrik, Charles Spence, and Richard E. Passingham. 2004. That's my hand! Activity in the premotor cortex reflects feeling of ownership of a limb. *Science* 305 (5685): 875–77.

Ehrsson, H. Henrik, Nicholas P. Holmes, and Richard E. Passingham. 2005. Touching a rubber hand: Feeling of body ownership is associated with activity in multisensory brain areas. *Journal of Neuroscience* 25 (45): 10564–73.

Ellis, Havelock. 1898. Mescal: A new artificial paradise. *Contemporary Review* 73: 130–41 (reprinted in the Smithsonian Institution Annual Report 1898, pp. 537–48).

Escher, Sandra, and Marius Romme. 2012. The hearing voices movement. In *Hallucinations: Research and Practice,* ed. Jan Dirk Blom and Iris E. C. Sommer. New York: Springer.

Fénelon, Gilles, Florence Mahieux, Renaud Huon, and Marc Ziégler. 2000. Hallucinations in Parkinson's disease: Prevalence, phenomenology and risk factors. *Brain* 123 (4): 733–45.

ffytche, Dominic H. 2007. Visual hallucinatory syndromes: Past, present, and future. *Dialogues in Clinical Neuroscience* 9: 173–89.
———. 2008. The hodology of hallucinations. *Cortex* 44: 1067–83.

ffytche, D. H., R. J. Howard, M. J. Brammer, A. David, P. Woodruff, and S. Williams. 1998. The anatomy of conscious vision: An fMRI study of visual hallucinations. *Nature Neuroscience* 1 (8): 738–42.

Foote-Smith, Elizabeth, and Lydia Bayne. 1991. Joan of Arc. *Epilepsia* 32 (6): 810–15.

Freud, Sigmund. 1891/1953. *On Aphasia: A Critical Study.* Oxford: International Universities Press.

———. 1901/1990. *The Psychopathology of Everyday Life.* New York: Norton.

Freud, Sigmund, and Josef Breuer. 1895/1991. *Studies on Hysteria.* New York: Penguin.

Friedman, Diane Broadbent. 2008. *A Matter of Life and Death: The Brain Revealed by the Mind of Michael Powell.* Bloomington, IN: AuthorHouse.

Fuller, G. N., and R. J. Guiloff. 1987. Migrainous olfactory hallucinations. *Journal of Neurology, Neurosurgery and Psychiatry* 50: 1688–90.

Fuller, John Grant. 1968. *The Day of St. Anthony's Fire.* New York: Macmillan.

Funk, Marion, Maggie Shiffrar, and Peter Brugger. Hand movement observation by individuals born without hands: Phantom limb experience constrains visual limb perception. *Experimental Brain Research* 164 (3): 341–46.

Galton, Francis. 1883. *Inquiries into Human Faculty.* London: Macmillan.

Gastaut, Henri, and Benjamin G. Zifkin. 1984. Ictal visual hallucinations of numerals. *Neurology* 34 (7): 950–53.

Gélineau, J. B. E. 1880. De la narcolepsie. *Gazette des hôpitaux* 54: 635–37.

Geschwind, Norman. 1984. Dostoievsky's epilepsy. In *Psychiatric Aspects of Epilepsy,* ed. Dietrich Blumer (pp. 325–33). Washington, D.C.: American Psychiatric Press.

———. 2009. Personality changes in temporal lobe epilepsy. *Epilepsy & Behavior* 15: 425–33.

Gilbert, Martin. 1997. *The Boys: The Story of 732 Young Concentration Camp Survivors.* New York: Holt.

Gowers, W. R. 1881. *Epilepsy and Other Chronic Convulsive Diseases: Their Causes, Symptoms and Treatment.* London: Churchill.

———. 1907. *The Border-land of Epilepsy.* London: Churchill.

Green, Celia. 1968. *Out-of-the-Body Experiences.* Oxford: Institute of Psychophysical Research.

Gurney, Edmund, F. W. H. Myers, and Frank Podmore. 1886. *Phantasms of the Living.* London: Trubner & Co.

Hayes, Bill. 2001. *Sleep Demons: An Insomniac's Memoir.* New York: Washington Square.

Hayter, Alethea. 1998. *Opium and the Romantic Imagination: Addiction and Creativity in De Quincey, Coleridge, Baudelaire and Others.* New York: HarperCollins.

Heins, Terry, A. Gray, and M. Tennant. 1990. Persisting hallucinations following childhood sexual abuse. *Australian and New Zealand Journal of Psychiatry* 24: 561–65.

Hobson, Allan. 1999. *Dreaming as Delirium: How the Brain Goes Out of Its Mind.* Cambridge, MA: MIT Press.

Holmes, Douglas S., and Louis W. Tinnin. 1995. The problem of auditory hallucinations in combat PTSD. *Traumatology* 1 (2): 1–7.

Hughes, Robert. 2006. *Goya.* New York: Knopf.

Hustvedt, Siri. 2008. Lifting, lights, and little people. In *Migraines: Perspectives on a Headache* (blog). *New York Times,* February 17, 2008. http://migraine.blogs.nytimes.com/2008/02/17/lifting-lights-and-little-people/.

Huxley, Aldous. 1952. *The Devils of Loudon.* London: Chatto & Windus.

———. 1954. *"The Doors of Perception"* and *"Heaven and Hell."* New York: Harper & Row.

Jackson, John Hughlings. 1925. *Neurological Fragments.* London: Oxford Medical.

———. 1932. *Selected Writings.* Vol. 2, ed. James Taylor, Gordon Holmes, and F. M. R. Walshe. London: Hodder and Stoughton.

Jackson, John Hughlings, and W. S. Colman. 1898. Case of epilepsy with tasting movements and "dreamy state"—very small patch of softening in the left uncinate gyrus. *Brain* 21 (4): 580–90.

Jaffe, Ruth. 1968. Dissociative phenomena in former concentration camp inmates. *International Journal of Psycho-Analysis* 49: 310–12.

James, William. 1887. The consciousness of lost limbs. *Proceedings of the American Society for Psychical Research* 1 (3): 249–58.

———. 1890. *The Principles of Psychology.* London: Macmillan.

———. 1896/1984. *William James on Exceptional Mental States: The 1896 Lowell Lectures,* ed. Eugene Taylor. Amherst: University of Massachusetts Press.

———. 1902. *The Varieties of Religious Experience: A Study in Human Nature.* London: Longmans, Green.

Jaynes, Julian. 1976. *The Origin of Consciousness in the Breakdown of the Bicameral Mind.* New York: Houghton Mifflin.

Jones, Ernest. 1951. *On the Nightmare.* New York: Grove Press.

Kaplan, Fred. 1992. *Henry James: The Imagination of Genius.* Baltimore: Johns Hopkins University Press.

Keynes, John Maynard. 1949. *Two Memoirs: "Dr. Melchior, a Defeated Enemy" and "My Early Beliefs."* London: Rupert Hart-Davis.

Klüver, Heinrich. 1928. *Mescal: The "Divine" Plant and Its Psychological Effects.* London: Kegan Paul, Trench, Trübner.

———. 1942. Mechanisms of hallucinations. In *Studies in Personality,* ed. Q. McNemar and M. A. Merrill (pp. 175–207). New York: McGraw-Hill.

Kraepelin, Emil. 1904. *Lectures on Clinical Psychiatry.* New York: William Wood.

La Barre, Weston. 1975. Anthropological perspectives on hallucination and hallucinogens. In *Hallucinations: Behavior, Experience, and Theory,* ed. R. K. Siegel and L. J. West (pp. 9–52). New York: John Wiley & Sons.

Lance, James. 1976. Simple formed hallucinations confined to the area of a specific visual field defect. *Brain* 99 (4): 719–34.

Landis, Basile N., and Pierre R. Burkhard. 2008. Phantosmias and Parkinson disease. *Archives of Neurology* 65 (9): 1237–39.

Leaning, F. E. 1925. An introductory study of hypnagogic phenomena. *Proceedings of the Society for Psychical Research* 35: 289–409.

Leiderman, Herbert, Jack H. Mendelson, Donald Wexler, and Philip Solomon. 1958. Sensory deprivation: Clinical aspects. *Archives of Internal Medicine* 101: 389–96.

Leudar, Ivan, and Philip Thomas. 2000. *Voices of Reason, Voices of Madness: Studies of Verbal Hallucinations.* London: Routledge.

Lewin, Louis. 1886/1964. *Phantastica: Narcotic and Stimulating Drugs.* London: Routledge & Kegan Paul.

Lhermitte, Jean. 1922. Syndrome de la calotte du pédoncule cerebral: Les troubles psycho-sensoriels dans les lésions du mésocéphale. *Revue Neurologique* (Paris) 38: 1359–65.

———. 1951. Visual hallucinations of the self. *British Medical Journal* 1 (4704): 431–34.

Lippman, Caro W. 1952. Certain hallucinations peculiar to migraine. *Journal of Nervous and Mental Disease* 116 (4): 346–51.

Liveing, Edward. 1873. *On Megrim, Sick-Headache, and Some Allied Disorders: A Contribution to the Pathology of Nerve-Storms.* London: J. & A. Churchill.

Luhrmann, T. M. 2012. *When God Talks Back: Understanding the American Evangelical Relationship with God.* New York: Knopf.

Macnish, Robert. 1834. *The Philosophy of Sleep.* New York: D. Appleton.

Maupassant, Guy de. 1903. *Short Stories of the Tragedy and Comedy of Life.* Akron, OH: St. Dunstan Society.

Maury, Louis Ferdinand Alfred. 1848. Des hallucinations hypnago-giques, ou des erreurs des sens dans l'état intermediaire entre la veille et le sommeil. *Annales medico-psychologiques du système nerveux* 11: 26–40.

Mavromatis, Andreas. 1991. *Hypnagogia: The Unique State of Consciousness Between Wakefulness and Sleep.* London: Routledge.

Mayeux, Richard, and D. Frank Benson. Phantom limb and multiple sclerosis. *Neurology* 29: 724–26.

McGinn, Colin. 2006. *Mindsight: Image, Dream, Meaning.* Cambridge, MA: Harvard University Press.

McKellar, Peter, and Lorna Simpson. 1954. Between wakefulness and sleep: Hypnagogic imagery. *British Journal of Psychology* 45 (4): 266–76.

Melville, Herman. 1851. *Moby-Dick; or, The Whale.* New York: Harper and Brothers.

Merabet, Lotfi B., Denise Maguire, Aisling Warde, Karin Alterescu, Robert Stickgold, and Alvaro Pascual-Leone. 2004. Visual hallucinations during prolonged blindfolding in sighted subjects. *Journal of Neuro-Ophthalmology* 24 (2): 109–13.

Merzenich, Michael. 1998. Long-term change of mind. *Science* 282 (5391): 1062–63.

Mitchell, Silas Weir. 1866. The case of George Dedlow. *Atlantic Monthly.*
———. 1872/1965. *Injuries of Nerves and Their Consequences.* New York: Dover.
———. 1896. Remarks on the effects of *Anhelonium lewinii* (the mescal button). *British Medical Journal* 2 (1875): 1624–29.

Mitchell, Silas Weir, William Williams Keen, and George Read Morehouse. 1864. *Reflex Paralysis.* Washington, D.C.: Surgeon General's Office.

Mogk, Lylas G., and Marja Mogk. 2003. *Macular Degeneration: The Complete Guide to Saving and Maximizing Your Sight.* New York: Ballantine Books.

Mogk, Lylas G., Anne Riddering, David Dahl, Cathy Bruce, and Shannon Brafford. 2000. Charles Bonnet syndrome in adults with visual impairments from age-related macular degeneration. In *Vision Rehabilitation (Assessment, Intervention and Outcomes)*, ed. Cynthia Stuen et al. (pp. 117–19). Downingtown, PA: Swets and Zeitlinger.

Moody, Raymond A. 1975. *Life After Life: The Investigation of a Phenomenon—Survival of Bodily Death.* Atlanta: Mockingbird Books.

Moreau, Jacques Joseph. 1845/1973. *Hashish and Mental Illness.* New York: Raven Press.

Myers, F. W. H. 1903. *Human Personality and Its Survival of Bodily Death.* London: Longmans, Green.

Nabokov, Vladimir. 1966. *Speak, Memory: An Autobiography Revisited.* New York: McGraw-Hill.

Nasrallah, Henry A. 1985. The unintegrated right cerebral hemispheric consciousness as alien intruder: A possible mechanism for Schneiderian delusions in schizophrenia. *Comprehensive Psychiatry* 26 (3): 273–82.

Nelson, Kevin. 2011. *The Spiritual Doorway in the Brain: A Neurologist's Search for the God Experience.* New York: Dutton.

Newberg, Andrew B., Nancy Wintering, Mark R. Waldman, Daniel Amen, Dharma S. Khalsa, and Abass Alavi. 2010. Cerebral blood flow differences between long-term meditators and non-meditators. *Consciousness and Cognition* 19 (4): 899–905.

Omalu, Bennet, Jennifer L. Hammers, Julian Bailes, Ronald L. Hamilton, M. Ilyas Kamboh, Garrett Webster, and Robert P. Fitzsimmons. 2011. Chronic traumatic encephalopathy in an Iraqi war veteran with posttraumatic stress disorder who committed suicide. *Neurosurgical Focus* 31 (5): E3.

Otten, Erna. 1992. Phantom limbs [letter to the editor and reply from Oliver Sacks]. *New York Review of Books* 39 (3): 45–46.

Parkinson, James. 1817. *An Essay on the Shaking Palsy.* London: Whittingham and Bowland.

Penfield, Wilder, and Phanor Perot. 1963. The brain's record of auditory and visual experience. *Brain* 86 (4): 596–696.

Peters, J. C. 1853. *A Treatise on Headache.* New York: William Radde.

Podoll, Klaus, and Derek Robinson. 2008. *Migraine Art: The Migraine Experience from Within.* Berkeley, CA: North Atlantic Books.

Poe, Edgar Allan. 1902. *The Complete Works of Edgar Allan Poe.* New York: G. P. Putnam's Sons.

Poeck, K. 1964. Phantoms following amputation in early childhood and in congenital absence of limbs. *Cortex* 1 (3): 269–74.

Ramachandran, V. S. 2012. *The Tell-Tale Brain.* New York: W. W. Norton.

Ramachandran, V. S., and W. Hirstein. 1998. The perception of phantom limbs. *Brain.* 121(9): 1603–30.

Rees, W. Dewi. 1971. The hallucinations of widowhood. *British Medical Journal* 4: 37–41.

Richards, Whitman. 1971. The fortification illusions of migraines. *Scientific American* 224 (5): 88–96.

Riddoch, George. 1941. Phantom limbs and body shape. *Brain* 4 (4): 197–222.

Rosenhan, D. L. 1973. On being sane in insane places. *Science* 179 (4070): 250–58.

Sacks, Oliver. 1970. *Migraine.* Berkeley: University of California Press.

———. 1973. *Awakenings.* New York: Doubleday.

———. 1984. *A Leg to Stand On.* New York: Summit Books.

———. 1985. *The Man Who Mistook His Wife for a Hat.* New York: Summit Books.

———. 1992. Phantom faces. *British Medical Journal* 304: 364.

———. 1995. *An Anthropologist on Mars.* New York: Knopf.

———. 1996. *The Island of the Colorblind.* New York: Knopf.

———. 2004. In the river of consciousness. *New York Review of Books,* January 15, 2004.

———. 2004. Speed. *New Yorker,* August 23, 2004, 60–69.

———. 2007. *Musicophilia: Tales of Music and the Brain.* New York: Knopf.

———. 2010. *The Mind's Eye.* New York: Knopf.

Salzman, Mark. 2000. *Lying Awake.* New York: Knopf.

Santhouse, A. M., R. J. Howard, and D. H. ffytche. 2000. Visual hallucinatory syndromes and the anatomy of the visual brain. *Brain* 123: 2055–64.

Scatena, Paul. 1990. Phantom representations of congenitally absent limbs. *Perceptual and Motor Skills* 70: 1227–32.

Schneck, J. M. S. 1989. Weir Mitchell's visual hallucinations as a grief reaction. *American Journal of Psychiatry* 146 (3): 409.

Schultes, Richard Evans, and Albert Hofmann. 1992. *Plants of the Gods: Their Sacred, Healing and Hallucinogenic Powers.* Rochester, VT: Healing Arts Press.

Shanon, Benny. 2002. *The Antipodes of the Mind: Charting the Phenomenology of the Ayahuasca Experience.* Oxford: Oxford University Press.

Shengold, Leonard. 2006. *Haunted by Parents.* New Haven: Yale University Press.

Shermer, Michael. 2005. Abducted! *Scientific American* 292: 34.

———. 2011. *The Believing Brain: From Ghosts and Gods to Politics and Conspiracies—How We Construct Beliefs and Reinforce Them as Truths.* New York: Times Books.

Shively, Sharon B., and Daniel P. Perl. 2012. Traumatic brain injury, shell shock, and posttraumatic stress disorder in the military—past, present, and future. *Journal of Head Trauma Rehabilitation*, in press.

Siegel, Ronald K. 1977. Hallucinations. *Scientific American* 237 (4): 132–40.

———. 1984. Hostage hallucinations: Visual imagery induced by isolation and life-threatening stress. *Journal of Nervous and Mental Disease* 172 (5): 264–72.

Siegel, Ronald K., and Murray E. Jarvik. 1975. Drug-induced hallucinations in animals and man. In *Hallucinations: Behavior, Experience, and Theory,* ed. R. K. Siegel and L. J. West (pp. 81–162). New York: John Wiley & Sons.

Siegel, Ronald K., and Louis Jolyon West. 1975. *Hallucinations: Behavior, Experience, and Theory.* New York: John Wiley & Sons.

Simpson, Joe. 1988. *Touching the Void.* New York: HarperCollins.

Sireteanu, Ruxandra, Viola Oertel, Harald Mohr, David Linden, and Wolf Singer. 2008. Graphical illustration and functional neuroimaging of visual hallucinations during prolonged blindfolding: A comparison to visual imagery. *Perception* 37: 1805–21.

Smith, Daniel B. 2007. *Muses, Madmen, and Prophets: Hearing Voices and the Borders of Sanity.* New York: Penguin.

Society for Psychical Research. 1894. Report on the census of hallucinations. *Proceedings of the Society for Psychical Research* 10: 25–422.

Spinoza, Benedict. 1883/1955. *On the Improvement of the Understanding, The Ethics, and Correspondence.* Vol. 2. New York: Dover.

Stevens, Jay. 1998. *Storming Heaven: LSD and the American Dream.* New York: Grove.

Strindberg, August. 1898/1962. *Inferno.* London: Hutchinson.

Swartz, Barbara E., and John C. M. Brust. 1984. Anton's syndrome accompanying withdrawal hallucinosis in a blind alcoholic. *Neurology* 34 (7): 969.

Swash, Michael. 1979. Visual perseveration in temporal lobe epilepsy. *Journal of Neurology, Neurosurgery, and Psychiatry* 42(6): 569–71.

Taylor, David C., and Susan M. Marsh. 1980. Hughlings Jackson's Dr Z: The paradigm of temporal lobe epilepsy revealed. *Journal of Neurology, Neurosurgery, and Psychiatry* 43: 758–67.

Teunisse, Robert J., F. G. Zitman, J. R. M. Cruysberg, W. H. L. Hoefnagels, and A. L. M. Verbeek. 1996. Visual hallucinations in psychologically normal people: Charles Bonnet's syndrome. *Lancet* 347 (9004): 794–97.

Thorpy, Michael J., and Jan Yager. 2001. *The Encyclopedia of Sleep and Sleep Disorders.* 2nd ed. New York: Facts on File.

Van Bogaert, Ludo. 1927. Peduncular hallucinosis. *Revue neurologique.* 47: 608–17.

Vygotsky, L. S. 1962. *Thought and Language,* ed. Eugenia Hanfmann and Gertrude Vahar. Cambridge, MA: MIT Press and John Wiley & Sons. Original Russian edition published in 1934.

Watkins, John. 1998. *Hearing Voices: A Common Human Experience.* Melbourne: Hill of Content.

Waugh, Evelyn. 1957. *The Ordeal of Gilbert Pinfold.* Boston: Little, Brown.

Weissman, Judith. 1993. *Of Two Minds: Poets Who Hear Voices.* Hanover, NH: Wesleyan University Press / University Press of New England.

Wells, H. G. 1927. *The Short Stories of H. G. Wells.* London: Ernest Benn.

West, L. Jolyon, ed. 1962. *Hallucinations.* New York: Grune & Stratton.

Wigan, A. L. 1844. *A New View of Insanity: The Duality of the Mind Provided by the Structure, Functions, and Diseases of the Brain.* London: Longman, Brown, Green, and Longmans.

Wilson, Edmund. 1990. *Upstate: Records and Recollections of Northern New York.* Syracuse: Syracuse University Press.

Wilson, S. A. Kinnier. 1940. *Neurology.* London: Edward Arnold.

Wittgenstein, Ludwig. 1975. *On Certainty.* Malden, MA: Blackwell.

Zamboni, Giovanna, Carla Budriesi, and Paolo Nichelli. 2005. "Seeing oneself": A case of autoscopy. *Neurocase* 11 (3): 212–15.

Zubek, John P., ed. 1969. *Sensory Deprivation: Fifteen Years of Research.* New York: Meredith.

Zubek, John P., Dolores Pushkar, Wilma Sansom, and J. Gowing. 1961. Perceptual changes after prolonged sensory isolation (darkness and silence). *Canadian Journal of Psychology* 15 (2): 83–100.

# Index

# Permissions Acknowledgments

Grateful acknowledgment is made to the following for permission to reprint previously published material:

American Academy of Neurology: Excerpt from "Anton's Syndrome Accompanying Withdrawal Hallucinosis in a Blind Alcoholic" by Barbara E. Swartz and John C. M. Brust from *Neurology 34* (1984). Reprinted by permission of the American Academy of Neurology as administered by Wolters Kluwer Health Medical Research.

American Psychiatric Publishing: Excerpt from "Weir Mitchell's Visual Hallucinations as a Grief Reaction" by Jerome S. Schneck, M.D., from *American Journal of Psychiatry* (1989). Copyright © 1989 by *American Journal of Psychiatry*. Reprinted by permission of American Psychiatric Publishing a division of American Psychiatric Association.

BMJ Publishing Group Ltd.: Excerpt from "Heautoscopy, Epilepsy and Suicide" by P. Brugger, R. Agosti, M. Regard, H. G. Wieser and T. Landis from *Journal of Neurology, Neurosurgery and Psychiatry*, July 1, 1994. Reprinted by permission of BMJ Publishing Group Ltd. as administered by the Copyright Clearance Center.

Cambridge University Press: Excerpts from *Disturbances of the Mind* by Douwe Draaisma, translated by Barbara Fasting. Copyright © 2006 by Douwe Draaisma. Reprinted by permission of Cambridge University Press.

Canadian Psychological Association: Excerpt from "Effects of Decreased Variation of the Sensory Environment" by W. H. Bexton, W. Heron and T. H. Scott from *Canadian Psychology* (1954). Copyright © 1954 by Canadian Psychological Association. Excerpt from "Perceptual Changes after Prolonged Sensory Isolation (Darkness and Silence)" by John P. Zubek, Dolores Pushkar, Wilma Sansom and J. Gowing from

A NOTE ON THE TYPE

The text of this book was composed in Trump Mediaeval. Designed by Professor Georg Trump (1896–1985) in the mid-1950s, Trump Mediaeval was cut and cast by the C. E. Weber Type Foundry of Stuttgart, Germany. The roman letterforms are based on classical prototypes, but Professor Trump has imbued them with his own unmistakable style. The italic letterforms, unlike those of so many other typefaces, are closely related to their roman counterparts. The result is a truly contemporary type, notable for both its legibility and its versatility.

Composed by North Market Street Graphics,
Lancaster, Pennsylvania

Printed and bound by RR Donnelley,
Harrisonburg, Virginia

Designed by Iris Weinstein